The Green Belt, Housing Crises and Planning Systems

This book evaluates the effectiveness of the Green Belt planning policy in England. It is one of the most well-known strategies internationally, with similar growth restraint policies having been adopted in a diverse range of cities around the world, such as Portland, Medellin and Bangkok. Despite this, it is often argued that Green Belts contribute to wider inequitable outcomes in society.

Focusing on the Green Belt in England, the book critically analyses the extent to which these policies and planning systems contribute to housing crises and examines how far they need to be reformed. With the central role of community engagement in many of the debates about housing crises, the book investigates the characteristics of popular and campaigner opposition to housebuilding alongside investigating the power relations and politics of planning systems. This timely research taps into important current policy debates in the UK surrounding urban nature, green infrastructure, and building on the 'grey belt'.

The book is therefore of relevance and benefit to policymakers and politicians, to academics and students internationally from a range of fields interested in housing, community engagement, green infrastructure, strategic planning, power and politics, and conservation.

Charles Edward Goode is an Assistant Professor in Urban and Regional Planning in the School of Geography at the University of Birmingham, UK. He is a geographer and trained planner with research and teaching interests in strategic planning and regional governance, community involvement, housing supply/affordability and planning history.

Routledge Research in Planning and Urban Design

Routledge Research in Planning and Urban Design is a series of academic monographs for scholars working in these disciplines and the overlaps between them. Building on Routledge's history of academic rigour and cutting-edge research, the series contributes to the rapidly expanding literature in all areas of planning and urban design.

Australia and China Perspectives on Urban Regeneration and Rural Revitalization
Raffaele Pernice and Bing Chen

Contested Airport Land
Social-Spatial Transformation and Environmental Injustice in Asia and Africa
Edited by Irit Ittner, Sneha Sharma, Isaac Khambule and Hanna Geschewski

Rebuilding Urban Complexity
A Configurational Approach to Postindustrial Cities
Francesca Froy

The Protection of Green Spaces for Climate Change Adaptation
Planning Systems, Policies and Instruments
Edited by Maciej J. Nowak

Exclusion in Smart Cities
Assessment and Strategy for Strengthening Resilience
Izabela Jonek-Kowalska and Radosław Wolniak

The Green Belt, Housing Crises and Planning Systems
Charles Edward Goode

For more information about this series, please visit: www.routledge.com/
Routledge-Research-in-Planning-and-Urban-Design/book-series/RRPUD

The Green Belt, Housing Crises and Planning Systems

Charles Edward Goode

Routledge
Taylor & Francis Group

LONDON AND NEW YORK

Designed cover image: Charles Goode.

First published 2025
by Routledge
4 Park Square, Milton Park, Abingdon, Oxon OX14 4RN

and by Routledge
605 Third Avenue, New York, NY 10158

Routledge is an imprint of the Taylor & Francis Group, an informa business

British Library Cataloguing-in-Publication Data
A catalogue record for this book is available from the British Library

ISBN: 978-1-032-67427-8 (hbk)
ISBN: 978-1-032-67428-5 (pbk)
ISBN: 978-1-032-67431-5 (ebk)

DOI: 10.4324/9781032674315

Typeset in Times New Roman
by KnowledgeWorks Global Ltd.

Dedicated to the loving memory of my dear Mother, Rachel Margaret Rose Goode, who fell asleep through Jesus on 26th November 2019, aged 57 (Song of Songs 2:17, 'Until the day dawn, and the shadows flee away'; Revelation 22:20, 'Yea, I come quickly. Amen; come, Lord Jesus').

Contents

Figures and tables

Figures

Tables

Preface

The Green Belt is probably the most well-known planning policy internationally, having been adopted around the world in cities such as Portland, Medellin and Toronto. It is associated with planning 'heroes', especially Ebenezer Howard and Sir Patrick Abercrombie. However, in common with wider critiques of planning systems internationally and following Peter Hall's seminal work *The Containment of Urban England* (1973), it is often argued that Green Belts (or Urban Growth Boundaries) alongside broader planning regulation contribute to wider inequitable outcomes in society. The core argument is that, through restricting the land supply for residential development, which raises house prices, planning systems are disproportionately skewed in favour of anti-development campaigners and homeowners. This juxtaposition is particularly prominent in England with the Green Belt being the most popular and longstanding planning policy. The policy commands widespread popular political support at the local and national levels, yet it is regularly critiqued as one of the main causes of England's housing crisis by academics, think tanks and housebuilders. For example, the Prime Minister and Leader of the Labour Party, Sir Keir Starmer has argued for building on the so-called 'grey belt' in the Green Belt whilst the formerly governing Conservative Party promised to protect the Green Belt – the policy was therefore a crucial issue in the 2024 General Election (Martin, 2024; Mason, 2023).

This book, firstly, evaluates the effectiveness of the Green Belt, which is very important given the policy's significance both in England and internationally, and there is the currency of wider economic arguments critiquing planning systems around the world. Secondly, this book critically analyses the extent to which the policy, Urban Growth Boundaries and planning systems more broadly contribute to housing crises, which is a pressing policy debate in many countries. Focusing on the Green Belt in England, the book examines how far the policy needs to be reformed, which is critical given the existence of similar policies around the world. Thirdly, with the central role of community involvement in many of the debates about housing crises, the book investigates the motivations and characteristics of popular and campaigner support for the policy and opposition to housebuilding generally.

The originality of the book is in addressing the significant research gap of examining these three core interrelated themes – Green Belt, housing crises and the politics of planning systems – and synthesising the findings in relation to the broader implications for theory and practice. There is a particular focus on the important, broader question of power in planning systems as there has arguably been a lack of geographically based case studies of power, especially of the rural-urban fringe, resulting in it being under-theorised in the literature. Drawing on an abundance of empirical material, this book underlines that there is a significant gap between the *attempted* and *effective* exercise of the power of campaigners in opposing housing development with significant circumscription and modulation of their power in the planning systems.

The book also has great relevance for theory and practice by advancing the often-polarised economic-conservationist debate on Green Belts and planning systems more broadly by focusing principally on the perspectives of planners as the main actors in the planning systems, yet their views are often under-represented in the Green Belt debate in England. This book explores the importance of space through a geographically based study of the West Midlands, therefore examining the housing crisis in under-researched post-industrial regions as there is a significant research gap with the dominance of first-tier cities in debates about housing supply. More broadly, the book has resonance with wider scholarly debates regarding green infrastructure and urban nature as demonstrating the outdatedness and challenge of responding to the Green Belt's modernist dualism and separation of rural and urban. However, the book examines the potentiality of a more environmentally focused, dynamic Green Belt policy moving forwards with its large spatial extent and proximity to most large cities and drawing upon the experience of alternative land-use policies in other countries. Indeed, the policy is at the centre of much debate about the future of land use, especially in the post-Brexit subsidy regime.

The book finds that the housing crisis in England is a complex, multi-faceted problem consisting of multi-scalar factors, although the Green Belt exacerbates the housing crisis in particular locations, especially on the edge of conurbations. In findings that have resonance with planning systems internationally, it argues that the policy should not be abolished but modified for the 21st century with a focus on sustainability, especially recreation and environmental improvement. Moreover, there should be a national conversation on the policy's overall spatial extent and purpose as well as its costs and benefits, perhaps as part of a broader national plan about patterns of housing and economic development in England. The study contributes to wider scholarship on community involvement in planning systems and opposition to new development by finding that people *primarily* support the policy because of popular planning principles and place attachment rather than house prices. Drawing on Lukes' three dimensions of power, the book highlights the circumscription and modulation of the power of campaigners opposing housing development in planning systems by planning policy and the development sector etc. Finally, many of the issues associated with the Green Belt and community opposition to development generally are related to the lack of strategic planning in the current system, so, given lessons which can be learned from practice internationally, the book underlines the need for integrated, strategic planning in England to protect the environment *and* meet housing need.

Acknowledgements

When thinking of the range of people who have helped and supported me with my PhD and this monograph, this section has proved very challenging to condense! First and foremost, many thanks to my excellent supervisors, Austin Barber and Mike Beazley, for their unwavering interest, care and support – this has been, is and will always be greatly appreciated. Additionally, thanks are due to the wonderful Planning Team at Birmingham, both past and present, including Kat Satler, Jeremy Whitehand (who sadly passed away in 2021), Adam Sheppard, Gethin Davison, Andrea Frank, Dave Adams, Katia Attuyer, Dilum Dissanayake and Rakib Akhtar (my former officemate who encouraged me to write the monograph). It is a real privilege being able to work alongside such excellent colleagues.

I would like to convey my appreciation to the wider School of Geography, Earth and Environmental Sciences at the University of Birmingham. Particular thanks to the until recently Head of School, Jon Oldfield, whose support, inspiration and mentorship have been greatly appreciated, and the new Head of School Nick Kettridge (whose passion for Hydrology I share with Physical Geography being weaved into the book's recommendations on an environmental purpose for the Green Belt!). Many thanks to my Geography colleagues, including Peter Kraftl, Sophie Hadfield-Hill, David Hannah, Julian Clark and Jessica Pykett. I am thankful for the wider Planning Academy and privileged to be part of a brilliant group of scholarly colleagues. Particular thanks to Charlotte Morphet, Quintin Bradley, Alan Mace, Gavin Parker, Nick Gallent and Andy Inch. I am also very grateful for the PGR community at Birmingham, who were a group of scholarly support and without whose guidance and sense of humour I would not have made it through. Phil Emerson, Alice Menzel, Katie Olivers, Stuart Bowles, Shivani Singh, Faye Shortland, Caz Russell, Amy Walker, George Willis, Rob Booth, Polly Jarman, Yanhui Shi, Lucy Capewell and others – I am truly grateful.

More broadly, the support of the planning profession, especially the RTPI West Midlands Regional Activities Committee and Young Planners Committee, is greatly valued and has helped to make this project. Particular thanks to Charlotte McGurk, Chris Young KC, Ellie Garrattley, Mike Best and Jonathan Manns for the very useful conversations. Finally, many thanks to all the undergraduate and postgraduate students I have taught in recent years. Whilst teaching responsibilities have made the writing of this book within time challenging, I thoroughly enjoy teaching and the debates and discussions that we have had have helped to shape and develop this book.

Of course, success in academia is dependent on the love and support of family and friends. Many thanks, particularly to my family, especially my Dad and sister, for your understanding, love and care. It is a great sorrow that my Mum is unable to witness the publication of this book, especially with the interest that she took in my work, but I trust that her legacy lives on. To my wider family and friends, many thanks for your interest, care and for putting up with me outside of work!

Finally, I hope that this book not only is useful in deepening our understanding of Green Belt and planning systems more broadly but that it also helps to shape the current Green Belt debate in policy and practice.

Charles Goode
Birmingham, February 2025

1 Introduction

Background

Green Belts, or Urban Growth Boundaries, are probably the most well-known planning policy internationally having been adopted in a diverse range of cities around the world such as Portland, Medellin and Toronto (Amati, 2007; Boyle and Mohamed, 2007). Indeed, given deepening concern about climate change mitigation and adaptation alongside management of urban growth, there is growing interest in Green Belts, particularly in the Global South (Chu *et al*, 2017). More broadly, the Green Belt is very important to a range of disciplines in both the Natural and Social Sciences, with the recent rapid growth in research on green infrastructure and urban nature as crucial to both climate change mitigation and urban well-being, particularly in a post-COVID context (Mell, 2024; Pykett, 2022; Reyes-Riveros *et al*, 2021). Indeed, the book engages with scholarly debate about moving to a more multi-functional Green Belt (see, for example, Kirby and Scott (2023) and Kirby *et al* (2023A, 2023B, 2024)).

In England, the Green Belt is the most well-known, iconic and long-standing planning policy, associated with internationally renowned planning 'heroes' and 'founding fathers' of the profession, such as Ebenezer Howard, Raymond Unwin and Sir Patrick Abercrombie, and implemented in the 1947 (Town and Country) Planning Act (Amati and Yokohari, 2007, p. 311). Whilst this book outlines some of the challenges of the policy being transferred internationally, such as in Tokyo and Sydney, in England the Green Belt has been very successful in its overall aim of shaping development patterns through effectively restricting urban 'sprawl' and commanding enduring and widespread political and popular support[1] (Department for Levelling Up, Housing and Communities (DLUHC), 2023, p. 42; Elson, 1986, p. 251).

However, in common with contemporary critiques of planning systems in many countries, the Green Belt is increasingly 'under attack' as a policy in the context of England's deepening housing crisis (Amati and Taylor, 2010, p. 143; Bradley, 2023; Goode, 2022A). For example, Prime Minister Sir Keir Starmer has argued for building on the so-called grey belt in the Green Belt, whilst the formerly governing Conservative Party promised to protect the Green Belt – it was therefore a crucial issue in the 2024 general election campaign (Martin, 2024;

DOI: 10.4324/9781032674315-1

Mason, 2023). The policy of the Labour Government on the Green Belt is still emerging, so this is acknowledged in the book rather than dwelt upon in depth, but with the Chancellor of the Exchequer, Rachel Reeves, promising a more 'strategic approach' on the Green Belt on 8th July 2024, there appears to be a significant change (Quinn, 2024, p. 1). Indeed, the significance of this book is that it helps to contextualise the current situation with Green Belt being a central political debate as, although both parties have historically been broadly in favour of the policy, the Green Belt is interwoven with broader political debates on localism and regionalism, the scope of community governance and power and approaches to solving the housing crisis.

Although the policy is criticised by the development sector and some planners (i.e. Baker, 2018; Griffiths, 2017), academic critiques stretch back to Peter Hall's (1973) seminal work, *The Containment of Urban England* (*Containment*), with attacks recently spearheaded and most critical by economists, especially Paul Cheshire (for example, Cheshire, 2024; Cheshire and Buyuklieva, 2019). There are similarities here with critiques of similar policies to Green Belts internationally, such as Urban Growth Boundaries in the USA and Australia (for example, Ball *et al*, 2014), as well as with broader planning systems and policies like heritage policies, zoning and environmental protection (for example, Hilber and Vermeulen, 2014). Indeed, Hall's (1973) *Containment* was a comparative study of the UK and the USA.

The principal argument of the 'economic' or 'land supply' school of reasoning is that the Green Belt restricts the land supply for residential development when and where demand for housing is highest at the rural-urban fringe, and therefore, it should be abolished or significantly reformed (with similar critiques employed of broader planning regulation internationally) (Cheshire, 2013, 2024; Hilber and Vermeulen, 2014). This book argues that disentangling the specific impacts of planning policies, like Green Belts, on house prices from other factors, such as locational characteristics, is challenging, especially with housing markets being characterised by spatial heterogeneity. It therefore calls for more geographically based case studies of housing crises to both inform theory and policy recommendations with the book empirically challenging propositions in the literature critiquing planning systems.

The Green Belt still enjoys widespread support in England and is often defended as sacrosanct with vigorous campaigns by groups nationally and locally, especially the Campaign to Protect Rural England (CPRE) and the Conservative Party as traditionally the party of the 'countryside' (Amati, 2008; CPRE, 2018). Again, in common with arguments in favour of planning systems internationally, the defence of the Green Belt centres on its environmental and sustainability value through restricting urban 'sprawl', whilst it is argued that the housing crisis is caused mainly by wider factors and can largely be solved through brownfield development (the 'conservationist argument') (Bradley, 2021, 2023; Bramley, 2019). Nonetheless, the book demonstrates that the Green Belt (and planning systems) do to some extent contribute towards housing crises, and therefore, it challenges the generalisations of the conservationist literature.

However, notwithstanding the Green Belt's prominence as a policy internationally and inherently geographical nature (in terms of its spatial *purpose* and widespread geographical *extent*), it remains under-researched and under-theorised in the academic literature (Elson, 1986; Mace, 2018; Sturzaker and Mell, 2016). Whilst some literature, especially Marxian, has focused largely on city centres and inner-cities and other literature on the rural, there has been a lack of spatial theorisation of the rural-urban fringe both in England and internationally (Kirby *et al*, 2023A, 2023B; Gallent *et al*, 2006; Scott *et al*, 2013). This is another major research 'gap' that this book aims to meet, especially as the Green Belt provides rich theoretical insights for conceptualising space and spatial governance. At the same time, drawing on rich empirical material, the book aims to move forward the broader debates on planning systems, which are often polarised, particularly the Green Belt with England's severe and deepening housing 'crisis'.

The book therefore has three interlocking and overarching themes:

- *The Green Belt:* an examination of the effectiveness of the policy, exploration of its potential reform including moving towards a more environmental, green infrastructure-based approach and analysis of the relationship between the policy and the housing crisis.
- *The housing crisis:* evaluation of the causation of the housing crises internationally, focusing on the Green Belt in England, examination of the possibilities of Green Belt reform and analysis of community opposition to housebuilding and support for the policy.
- *The politics of planning systems*: an exploration of the contribution of planning systems, especially the Green Belt policy, to unequal power relations in society, an examination of the modalities of the exercise of power in planning and an investigation of the politics of planning systems.

Shape and Scope

To understand the political dynamics of the Green Belt policy, the question must be started with – if the Green Belt is the principal cause of the housing crisis in England (the 'economic school' argument), why does it still command widespread popular and political support and has not been abolished or significantly reformed? This led the author to an interest in community opposition to housebuilding and the political nature of planning systems to find potential answers (Sturzaker, 2010). In this book, the author is particularly interested in the views of planners on the policy because they play a key role in formulating and implementing planning policy and, in particular, managing a *regional* growth management policy in a governance context of the 'localism agenda' in England (Allmendinger and Haughton, 2011, p. 317). However, in popular and academic discourse beyond the profession, they are often 'caught in the middle' in the Green Belt debate between 'economic' (i.e. significantly reform or abolish) and 'conservationist' (i.e. protect from any reform) arguments with their voices arguably under-represented in research (Mace, 2018; Goode, 2022A). The book is therefore a confluence of these academic, policy and practitioner interests with the Green Belt being a very relevant topic with crucial implications for policy and practice *and* an inherently spatial object of study, so it is ripe for theorising space. Indeed, the book aims to forward Hall's work through being case study and geographically based but also more theoretically focused. Moreover, given the similarities between the critiques of the Green Belt and planning systems more broadly in many countries (e.g. Hilber and Vermeulen, 2014), the book has broader implications for planning theory and practice.

Firstly, the study's golden thread is the importance of space, and as the Green Belt debate has often centred upon London and the South East of England which arguably has a distinctive housing market and transport infrastructure like other world cities such as New York and Tokyo (Edwards, 2015, 2016A; RTPI, 2016B), this book evaluates the effectiveness of *regional* Green Belts. The West Midlands is used as an 'exemplifying case' both temporally of the challenges of managing a strategic policy under the 'localism agenda' since 2010 but also geographically of many post-industrial regions around the world which have tight geographical restrictions or administrative boundaries (Allmendinger and Haughton, 2011, p. 317; Bryman, 2012, p. 51). More broadly, in relation to critiques of regional planning and Harrison *et al*'s (2021, p. 1) provocation that 'regional planning is dead', this book explores the potentiality and feasibility of more 'intermediate' or 'softer' spaces of governance building on the case study of the West Midlands (Valler and Phelps, 2018, p. 1).

Secondly, although the Green Belt debate is often ideological, even economic models struggle to disentangle its specific effects on house prices in common with other planning restrictions internationally (e.g. Bramley, 1993; Hilber and Vermeulen, 2014). Moreover, the increasing recognition of the socially constructed nature of housing markets justifies this book's more qualitative approach to housing and planning policy analysis (Munro, 2018). Consequently, its empirical backbone is the views of planners and planning stakeholders as experts to evaluate the (complex) relationship between the Green Belt and housing crisis and to help move the debate constructively forward.

Thirdly, many economic studies arguably do not take sufficient account of the political nature of planning systems, often recommending politically unfeasible proposals, such as abolishing or significantly reforming the Green Belt (Hilber and Vermeulen, 2014). However, the originality of this book derives from policy evaluation and recommendations which are cognisant of the politics of planning based upon current and extensive data collection as well as theoretical conceptualisation (Breheny, 1997; Cherry, 1982).

Fourthly, in contrast to practitioner studies, the research draws upon *theoretical insights* in analysing policy whilst simultaneously seeking to refine broader planning and geographical theory based upon the study's empirical data (Parker *et al*, 2015; Yeung, 1997, p. 520). It uses a critical realist and qualitatively led approach within an overall mixed-methods framework to recognise the spatial interconnectedness of the national, regional and local, especially in the centralised English planning system, alongside the need for triangulation because of the multi-faceted nature of the Green Belt and the housing crisis (Modell, 2009; Pike *et al*, 2018).

Research Propositions and Questions

The study developed theoretical propositions for empirical evaluation, but these propositions were grounded in the literature on policy analysis as well as planning and geographical theory as outlined in Chapter 2 (Marsh and McConnell, 2010; Palfrey *et al*, 2012). Firstly, working on the proposition that the Green Belt is the principal cause of the housing crisis (the economic school argument (Cheshire and Buyuklieva, 2019; Cheshire 2024)), it was hypothesised that planning systems, but particularly the Green Belt, are primarily supported by homeowners to maintain their property values. This drew upon and applied a political-economic framework whereby the Green Belt can be conceptualised as the ultimate arena of contestation in an 'intra-capitalist' conflict (Foglesong, 1986, p. 22; Hamiduddin and Gallent, 2024). This is between homeowners seeking to maximise their capital (house prices) through protecting the policy and housebuilders desiring a more flexible Green Belt to maximise their capital accumulation through (more) housebuilding (Gallent, 2019; Inch *et al*, 2020; Lake, 1993, p. 88). Moving theoretically 'upstream' (Soja, 2010), if this was the case, then outcomes from planning systems would be *significantly* skewed in favour of homeowners, and there are therefore uneven power relations in the planning system in England perpetuated by the Green Belt (Gallent *et al*, 2021; Rydin and Pennington, 2000; Short *et al*, 1987). Moving theoretically downstream, the study analyses *how* power is exercised by campaigners drawing principally on (and critiquing) Lukes' (2004, p. 14) *three dimensions of power* to elucidate the often more subtle ways that power is exercised at different spatial scales, particularly through discourse and imagination (see also Haugaard (2021)).

However, this overarching theoretical generalisation about power in planning depends upon the central research question about the extent to which planning systems, in this case the Green Belt, are the principal cause of housing crises and how far they need reforming. This question follows Hall's (1973, 1974) approach in *Containment* and other academic work on social and spatial justice, which analyses policy in terms of which groups 'gain' and 'lose' (Kiernan, 1983, p. 84; Sandercock and Dovey, 2002, p. 152). Moving from theory to practice, if the proposition was 'proved' regarding the Green Belt and planning system being the main cause of housing crises, developing recommendations on how the policy could be *significantly* reformed would be imperative, based on what is politically feasible and a recognition of the uneven power and social/spatial relations (Crook and Whitehead, 2019). The book therefore empirically explores and tests these crucial theoretical and policy propositions. Whilst focused upon England, the book has broader international relevance because of both the similarities between the Green Belt and other policies around the world, such as Urban Growth Boundaries, and the shared characteristics of critiques of the Green Belt and international critiques of planning systems.

Significance of the Research: Housing Crises, Power in Planning Systems and Green Infrastructure

Housing Crises

The focus of this book on housing is particularly poignant and relevant to policy because the concept of a housing 'crisis' is a long-standing but increasingly pressing 'wicked problem' in many cities and countries (Gallent *et al*, 2021; Christophers, 2018, 2019). It is argued to be a particularly an urban 'problem' linked to waves of gentrification from the rise of 'rentier' capitalism and the global circuits of capital, especially in globally attractive property 'hotspots', such as London and New York (Aalbers, 2019; Fernandez *et al*, 2016; Wetzstein, 2017).

More broadly, housing is the most significant driver of inter- and intragenerational inequality in England and at the top of the domestic political agenda with the Government's 'target' of 300,000 new homes annually (Lund, 2019; Martin, 2024, p. 36; Neuman, 2011, p. 154). As highlighted in Chapters 4 and 6, it is a multi-faceted problem, thus meriting the term '*housing crisis*' rather than just '*affordability crisis*' although it is centred on affordability with this book focusing on house prices (Gallent, 2019). It is argued that the high level and rising nature of house prices is increasingly pricing people, especially younger people, out of homeownership and forcing them into rented accommodation (Lund, 2019). For example, in 2017, in the UK, 30-year-olds were only half as likely to own their homes as 'baby boomers' at the same age, four times more likely to rent, and whilst homeowners during the 1960s/70s spent around 5% of their income on mortgages, private renters spent 36% (Watling and Breach, 2023,, p. 4; Corlett and Judge, 2017, pp. 4–5; Taylor, 2015, p. 38). The term 'housing crisis' is therefore used throughout the book as describing a serious problem which it is very important to resolve.

However, whilst particularly intense in England, the concept of the 'housing crisis' relates this book to the similar housing affordability issues in many countries around the world (Christophers, 2018, 2019). Firstly, there are urban areas with high demand for housing, especially first-tier cities such as New York, Christchurch or Vancouver, which has resulted in rising rents/house prices, international investor purchases of second homes and growing use of Airbnb and displacement of many people into surrounding, more affordable suburbs who commute back into urban cores (Aalbers, 2019; Bradley, 2023; Gallent *et al*, 2021; Wang *et al*, 2024). Likewise, many rural areas in close proximity to urban areas, such as Marin County (near San Francisco), Versailles (near Paris) or Lidingö (near Stockholm), are characterised by high house prices and second homeownership, especially in a post-COVID context (Hamiduddin and Gallent, 2024; Squires, 2022).

Secondly, there is a shortage of housing supply in places, such as the San Francisco Bay Area or Christchurch, caused by geographical or administrative restrictions (Squires, 2022), community opposition, especially in countries like the USA where residents can vote for restrictive zonal ordinances (Goode, 2022A) and the challenging politics of suburban densification, such as in Oslo (Kjærås,

2024). Thirdly, there have been global housing market booms and crashes which have been related to low interest rates and international financial markets (Wang *et al*, 2024; Wetzstein, 2017). Fourthly, in countries like the USA, there is the existence of popular political movements advocating for more housebuilding and intergenerational justice for younger people, exemplified in the global 'YIMBY' movement (Davis and Huennekens, 2022). Fifthly, in places where there are similar planning restrictions to the Green Belt, such as Urban Growth Boundaries in Portland and Melbourne, studies have identified these as being determinants in rising house prices, which has led to the abolition or loosening of these restrictions in places like Portland and Christchurch (Ball *et al*, 2014; Gallent *et al*, 2021; Layzer, 2012).

Research that seeks to solve housing crises is therefore very important, and there is wider geographical significance and relevance of the book's findings and recommendations in terms of 'lesson learning' for other countries (Squires and Heurkens, 2014, p. 2). Moreover, a central pillar of this book is that recommendations need to be politically and socially feasible alongside being theoretically derived and evidence-based due to the political nature of planning systems, so it has international relevance to policy as well as theory (Crook and Whitehead, 2019).

Community Involvement, Politics and Power in Planning Systems

The significance of this book also comes from directly relating an empirical study of the housing crisis to community involvement in and the politics of planning and, more broadly, power in planning systems and society. This not only connects the book to wider theoretical debates, but these themes of housing, community involvement, politics and power are strongly related to arguments made in the literature about campaigners opposing housing development primarily due to concern about house prices and often based on the assumption that campaigners have significant or disproportionate power to restrict housebuilding through effectively influencing the politics of planning (for example, Dehring *et al*, 2008; Rydin, 2020; Sturzaker, 2010). However, these arguments and assumptions, alongside the relationship between planning policy and the housing crisis, have arguably not been subject to sufficient empirical scrutiny, so this is a significant research gap that this book fills (Goode, 2022A). The book also forwards conceptual framings of the motives and characteristics of community opposition by underlining the importance of place attachment, environmental psychology and cultural geography, which have significant implications for practice and policy. Finally, alongside drawing upon and critiquing Lukes' (2004) *three dimensions of power* to elucidate the often more subtle ways that power is exercised at different spatial scales, the book advances the theoretical understanding of power in planning systems by highlighting that a significant gap exists between the *attempted* and *effective* exercise of power of campaigners in opposing housing development. Again, this has significant implications for public policy and planning practice.

Political Ecology/Green Infrastructure

The significance of the book additionally comes from exploring the relationship between the international debates on the future of land use, green infrastructure and the policy. On the one hand, there is a significant body of literature which demonstrates the benefits of green infrastructure and urban nature as crucial to both climate change mitigation and urban well-being, particularly in a post-COVID context (Mell, 2024; Pykett, 2022; Reyes-Riveros *et al*, 2021). On the other hand, there is the political ecology literature, which originated with studies in the Global South and explores the capitalisation and exploitation of nature and greenspace (for example, Malik, 2024; Walker, 2007). Although recreational access or environmental sustainability is not stated in the purposes of the Green Belt in England (Mace, 2018), the policy is very significant in broader debates about green infrastructure with the enduring challenge of it being constructed on the modernist dualism of separating rural and urban (Kirby and Scott, 2023; Kirby *et al*, 2023A, 2023B, 2024). Firstly, there are global debates about the *integration* of green infrastructure and built form as a way to both improve human physical and mental well-being and address climate change through, for example, mitigating Urban Heat Island and increasing water attenuation to reduce flooding. This is seen in urban farming and rewilding movements in many cities around the world. Secondly, there are debates about how land use in Urban Growth Boundaries can be moved towards the integration of rural and urban, especially in the Global South, with Green Belts introduced in Durban (South Africa) and Medellin (Colombia) to more sustainably manage land use and mitigate climate change, whilst Green Belts in Southeast Asia, in Bangkok (Thailand) for example, often include paddy fields to prevent flooding (Amati, 2008; Chu *et al*, 2017).

With the significant spatial extent of the Green Belt and its proximity to most large cities in England, the book examines the extensive debates about Urban Growth Boundaries and develops recommendations about the policy moving from a negative and passive policy to a more dynamic, environmentally focused one, especially in a post-Brexit subsidy regime. This would focus on rewilding/wildflower meadow planting, river restoration, urban farming and enhanced recreational access (see also Kirby and Scott's work on multi-functionality – Kirby and Scott (2023) and Kirby *et al* (2023A, 2023B, 2024)). The book therefore contributes to broader debates about urban nature, environmental justice and political ecology whilst drawing on international debates and examples of more environmentally focused land-use management policies to inform the Green Belt debate.

Structure

After this Introduction, Chapter 2 sets out the aims of the study and the mixed-methods approach used and introduces the main case study of the West Midlands Green Belt. There are two chapters (Chapters 3 and 4) on the history and form/function of the Green Belt before a 'bridge' chapter (Chapter 5) on conceptualising Green Belts, which draws upon the preceding chapters. Chapter 3 is on the Green

Belt's history and traces how it gained primacy as England's most long-standing and popular planning policy in contrast to other countries despite regular waves of deregulation and neo-liberalism (Mace, 2018; Prior and Raemaekers, 2007). Chapter 4 is analytical exploring the economic literature on the relationship between planning systems and housing affordability issues. It examines the Green Belt's purpose/objectives, spatial extent, international comparisons and alternative policies internationally. Given the paucity of theoretical research on the Green Belt, the conceptual frame chapter (Chapter 5) critically explores the different ways Green Belts can be theorised with a focus on the political-economic literature and power and politics in planning.

The second, empirical 'half' of the book is divided into four broad themes: Green Belt and the housing crisis (6), community opposition (7), protestors and politics (8) and governance (9). The housing Chapter (6) critically evaluates the extent to which the Green Belt is responsible for housing crises and explores its reform (covering practical, environmental and economic feasibility). The community Chapter (7) examines why campaigners and the public support the Green Belt and oppose housebuilding generally, thereby developing recommendations on how these concerns can be addressed (social feasibility). The politics Chapter (8) interrogates the crucial issue of power in planning, including how campaigners attempt to exercise power and the power of discourse and imagination in planning (political feasibility). These empirical chapters culminate in the governance Chapter (9) which, reflecting on practice internationally, develops recommendations on the need for strategic planning to address many of the issues raised in the book. These themed chapters draw out theoretical and policy/practitioner implications, based upon the data, and reflect Breheny's (1997, p. 209) argument on the importance of social and political feasibility in planning, so the book's recommendations have relevance for broader planning systems and housing policy generally. The conclusion chapter (Chapter 10) draws together the empirical chapters and develops the wider implications for theory and practice.

Note

1 The book focuses throughout on the Green Belt in England as those in Wales, Scotland and Northern Ireland function marginally differently due to political devolution (Lloyd and Peel, 2007; Warren and Clifford, 2005).

2 Researching the Green Belt

Introduction

This chapter forms the bridge between the introduction (Chapter 1) and the thematic chapters (Chapters 3–5) and empirical chapters (Chapters 6–9) by setting out the research questions framing this book and explaining the methods used, including introducing the case study of the West Midlands Green Belt. Building on the themes developed in Chapter 4, the research design explores the extent to which the Green Belt is the central cause of the housing crisis, whilst, as developed in Chapter 5, the design also aims to elucidate more broadly who has the most power in planning systems and how power is exercised. However, although planning is an inherently applied discipline (Goodman *et al*, 2017; Rydin, 1985), consideration of the epistemological and pragmatic aspects of research, especially methods, are still key (Bryman, 2012; Yeung, 1997). The chapter first sets out the research questions before exploring the main case study of the book. How these questions are operationalised, including the mixed-methods approach used, are then explained before outlining the methods themselves.

Research Questions

The overarching research aim, drawing together and summarising the main themes of the succeeding chapters (Chapters 3–5), is:

> To critically evaluate the relationship between the Green Belt, England's housing crisis and planning systems more broadly.

In common with most academic (planning) research, this book focuses on planning/geographical theory *and* planning policy/practice (Maidment, 2016; Rydin, 1985). It aims to conceptualise the Green Belt's relationship with the planning system and society, thereby exploring how the policy could be improved as theory and practice are deeply intertwined (Healey, 1997, 2003). Whereas many planning studies either focus on planning *processes* or *outcomes* (Lennon, 2020; Rydin, 1985), this study brings them *together* by focusing on

DOI: 10.4324/9781032674315-2

the Green Belt, leading to sequential research questions which operationalise the overall aim:

1   *To what extent is the Green Belt the principal cause of England's housing crisis?*

Given that the literature suggests it is *a* cause and the amount of public discourse on the Green Belt (Raynsford Review, 2018A), this justifies proceeding with this research project. The question is therefore more *'how much'* or *'to what extent'* is the policy the *main* cause of the housing crisis, so:

2   *Does the Green Belt need reform to solve the housing crisis? If so, to what extent? If not, what other policies are needed to solve the housing crisis?*

Further operationalising this question by the more theoretical analysis question, similar to that of Hall *et al* (1973A, 1973B):

3   *Who gains/benefits or loses/is disadvantaged from the Green Belt policy, and, crucially, to what extent are these gains/losses problematic to society? Does it reflect the underlying power structures of groups which have the most power in society?*

To what extent these gains and losses *are* problematic to society relates to the crucial questions based on planning's political nature and uneven power relations (Flyvbjerg, 1998, 2004):

4   *How and why are groups exercising power in relation to the Green Belt (i.e. why do people support it)?*

5   *More broadly, how can the Green Belt be conceptualised, and what are the implications for broader planning and geographical theory?*

Finally, this feeds back into the crucial policy question:

6   *To what extent is reforming the Green Belt realistically and practically feasible/ possible, especially in social and political terms (see Table 5.2 in Chapter 5)?*

Research Design

Having explored philosophical considerations, the research design operationalises this book's critical realist approach around triangulation and mixed-methods in relation to spatial scales and the main actors/groups in planning to 'measure' and explore the Green Belt from different spatial scales and stakeholder perspectives[1] (Neuman, 2011; Yeung, 1997). The main actors are planners (both in the private sector, working for developers/consultancies and the public sector), planning stakeholders and the public[2], especially campaigners, whilst the spatial scales are local, regional and national. Given this complexity and the range of factors at work, the project uses an 'integrative logic', which Mason (2006A, p. 6) defined as: 'mixing methods to ask questions about connecting parts, segments or layers of a social whole'. The 'social whole' is the Green Belt and its relationship with the housing crisis and the planning system, whilst the 'parts, segments or layers' are the key actors *and* spatial scales which are related to and joined together by the 'mechanisms' (governance, community opposition, housing and politics). The research instruments were designed by this integrative logic, whilst the methods themselves were developed and integrated to understand different actors and spatial scales through the mechanisms (Mason, 2006B, pp. 6–7, 2018).

Turning from 'how' methods were mixed to 'why', quantitative methods within a critical realist approach of not providing a *complete* representation of reality can still be useful as a means of establishing general trends and causation, the 'what', such as the views of planners and the public on the Green Belt (Field, 2013). This is used alongside qualitative methods, using an 'integrative logic', to explore the 'why' and responses to this (the 'how'), related to aspects of the study about human behaviour and policy, including the housing crisis and planning's political nature (Kvale, 2007; Onwuegbuzie and Leech, 2005). For example, Winchester (1999) used questionnaires alongside interviewing when researching lone fathers in Newcastle, Australia, using a similar rationale. The research therefore employs a pragmatic, qualitatively-led approach within an overall realist, mixed-methods framework as aiming to triangulate to find out about *as much as possible* about the Green Belt rather than trying to find a *representative, generalisable* sample (often associated with positivism) (Saunders *et al*, 2016).

The research methods are summarised below (Table 2.1):

Research Strategy: Methods

In translating the research design into a practical strategy, a unified core set of aims was used across different spatial scales and groups relating to the literature and key

Table 2.1 Research methods

Methods - National
- National: Planners - interviews and focus groups with key national planners and campaigners.
- National: General public - data analysis, i.e. CPRE Green Belt (GB) Survey, to explore popular attitudes.

Methods - Regional
- Regional: Planners - qualitative - interviews/focus groups with planners, mainly in the West Midlands. Quantitative - GB questionnaire of West Midlands.
- Regional: General public - case studies with GB campaign groups and interviews with campaigners.

mechanisms. This ensured that the aims and questions directly reflected the underlying policy and theoretical aims as recommended in the literature (Saunders *et al*, 2016).

Considering Spatial Scales

The strategy and design are anchored in this book around geographical scales, thereby reflecting the author's view as a geographer/planner on the importance of *space* as a shaper and shaped by social relations (Massey, 1994; Soja, 2010). Firstly, considering how the Green Belt is perceived and viewed *nationally* is important to frame and contextualise the study alongside allowing it to draw wider conclusions (May, 2011; Sturzaker, 2010; Yin, 2017). However, *purely* focusing nationally would miss the richness of spatial variation and place specificity regarding the Green Belt's largely local management (Mace, 2018). Additionally, it is important for theory and practice to evaluate its effectiveness *across* geographical areas to conceptualise the Green Belt more broadly in *regional* England (Edwards, 2015, 2016A; Kilroy, 2017). It was in this spatial framework that the regional case study design was developed.

Considering Planning Stakeholders

The views of two main groups can be identified: professional planners/planning stakeholders and the public, especially campaigners[3]. Consequently, the research *audience* was segmented at the two spatial scales, regional and national, and then the most appropriate methods were considered and chosen pragmatically to suit the research questions alongside epistemological commitments (Bryman, 2012; Yeung, 1997).

Merging Space and Stakeholders

Nationally, detailed data exists on the public's views of the Green Belt (CPRE's *Green Belt Questionnaire* (2015)) and housebuilding (*Social Attitudes Survey* (Park *et al*, 2012)) and interviews were initially conducted nationally with professional planners and planning stakeholders to contextualise the study.

Regionally, there is not an equivalent dataset on popular attitudes of the Green Belt and, given the cost and practicalities involved in sampling a general population, such as CPRE's Oxfordshire survey (Whall, 2015), alongside the author's qualitatively led approach, campaigners were focused on. This was to capture the views of a *specific* group, rather than the general population, but still look at 'professional' *and* 'everyday' campaigners in live regional campaign groups, as 'everyday' campaigners are an under-researched group (Amati, 2007; Bradley, 2019A, 2019B)[4].

In terms of planners, the author conducted a Green Belt questionnaire which was advertised in the RTPI West Midlands Magazine, *Tripwire*, and extensively on LinkedIn. Planners/planning stakeholders were interviewed regionally in the West Midlands alongside other regions (Yin, 2017).

Data Collection Methods: Case Study Approach

Case Study Approach Taken

The case study approach and the case itself (West Midlands Green Belt (WMGB)) reflect the overall research aims and the abductive, critical realist approach along-side spatial considerations (Silva *et al*, 2014; Yeung, 1997, p. 51)[5]. Given the Green Belt's vast spatial extent and time/resource constraints, the study had to focus on a small number of cases (Mason, 2018). Secondly, case study research aiming to explore the richness and 'thick' detail of the case(s) usually fits within an *overall* qualitative approach (Scapens, 2004, p. 275; Yin, 2017). However, conducting a series of 'mini' case studies in different parts of the UK was ini-tially envisaged with attempts at statistical generalisation, similar to Sturzaker's (2010) five case studies of power and planning in rural authorities[6]. Never-theless, the methodological literature highlights how this approach can 'lose' some of the richness and 'thick descriptions' of place that a small number of cases affords (Bryman, 2012, p. 52; Scapens, 2004; Tracy, 2010, p. 841). Con-sequently, it was decided that focusing in depth on a *single* case, the WMGB would result in a richer, fuller study with the variety of issues in the WMGB, whilst the dual focus on the national still broadened the study so that it has wider pol-icy relevance (Scapens, 2004; Yin, 2017). Moreover, the study evaluates the Green Belt's effectiveness *regionally* rather than just locally (Wannop and Cherry, 1994).

Case Study: West Midlands Green Belt – History and Current Issues

The WMGB covers nearly 225,000 ha and forms a continuous 'ring', between 5 and 7 miles wide, around the conurbation (which is home to nearly 2.9 million peo-ple) (Figure 2.1; CPRE and Natural England, 2010, p. 29). The WMGB was pro-posed in the West Midlands Group Study *Conurbation* (1948), tentatively agreed in 1955 but not formally approved until 1976 (Hall *et al*, 1973B, p. 584).

As Chapter 3 documents, Birmingham grew rapidly in the Industrial Revolu-tion, and suburban growth progressed apace in the inter-war period. This pros-perity continued in the post-war era, which, alongside post-war slum clearance, proposals for new towns and expanding existing towns outside the Green Belt, explains why the WMGB's boundaries were drawn so tightly[7] (Barber and Hall, 2008; Hall *et al*, 1973B, pp. 51–85; Law, 2000). Birmingham's acute housing needs were built during the 1950s, especially as it required an overspill popula-tion nearby as a local labour force (Sturzaker and Mell, 2016, p. 28). Like Man-chester's applications at Mobberley and Lymm, this culminated in the Wythall Inquiry (1958), whereby Birmingham's application to build a satellite settlement in the WMGB on the Worcestershire/Warwickshire boundary was refused by the Housing Minister due to it being in the Green Belt and widespread opposition by the rural neighbouring shires (see Chapter 9) (Figure 2.1; Cherry, 1996, p. 152). Eventually, Richard Crossman, the (Labour) Minister for Housing, approved Birmingham's application for an estate at Chelmsley Wood, Warwickshire, for 52,000 people in 1964[8] (Sturzaker and Mell, 2016, p. 28).

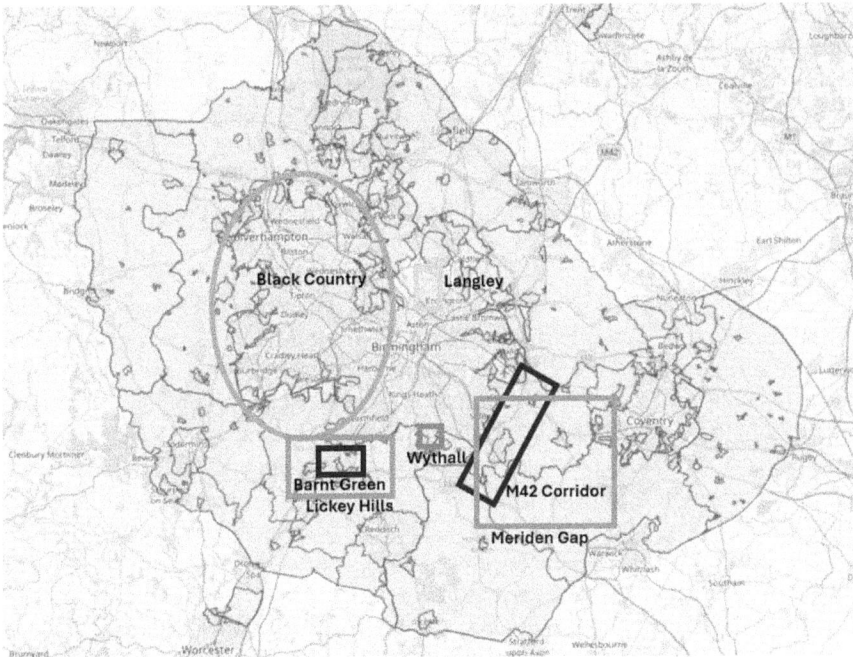

Figure 2.1 Key locations in the West Midlands (adapted from: https://bit.ly/3rXTGuW).

Key: M42 Corridor – an economically buoyant area on the rural-urban fringe of the West Midlands Conurbation.

Meriden Gap – the WMGB separates Birmingham and Coventry and is a very affluent area.

Black Country – a historically industrial area to the northwest of Birmingham composed of the Metropolitan Boroughs of Wolverhampton, Dudley, Sandwell and Walsall.

Wythall – a village on Birmingham's rural-urban fringe where the City Council applied to build an overspill estate

Barnt Green – a wealthy village in the WMGB near the Lickey Hills with a station on the Cross City line.

Lickey Hills – a range of hills on Birmingham's southern fringe. Most of the land is owned by the City Council and well-used for recreation by the conurbation.

Langley – a development of 6,000 homes adjoining Walmley, Sutton Coldfield.

The basis of the region's economic success, (over)reliance on the car industry, was the principal cause of its decline and deindustrialisation (Barber and Hall, 2008). From the 1980s, the primary focus of the region has shifted towards urban regeneration and brownfield development, with some Green Belt land released for employment, especially along the 'M42 Corridor', and very limited release for housing (Figure 2.1; Law, 2000, p. 64).

Birmingham is again the region's economic heart with a fast-growing, diverse population with a population growth of 150–200,000 by 2031 predicted in the Birmingham Development Plan (BDP) (from 2016) and a 'need' for 89,000 new houses (Birmingham City Council, 2017, p. 6). Consequently, there is huge development pressure on the WMGB, as Birmingham City Council claims to only have 'room' on brownfield land for 39,000 houses and released some of the extremely limited land in the WMGB within its boundary for 6,000 homes at Walmley (Figure 2.1; Best, 2019; Carpenter, 2016). The release was strongly opposed by the campaign group, Project Fields and constituency MP, Andrew Mitchell, who convinced the then Secretary of State, Greg Clark, to put a Holding Direction on the local plan (Best, 2019)[9]. This was subsequently lifted in 2016, with the government recognising Birmingham's housing needs (Goode, 2023).

Nonetheless, governance issues continue and are compounded because Birmingham City Council still claims to have a shortfall of 38,000 homes, which can only be met by neighbouring Local Planning Authorities (LPAs), which are also heavily constrained by the WMGB, such as the Black Country and Bromsgrove (Carpenter, 2016). Moreover, the absence of strategic planning and limited strength of the Duty to Cooperate (DtC)[10] on strategic matters between LPAs means that the WMGB is an excellent case study for exploring the challenges of the Green Belt's governance within the context of the 'localism agenda' and a pressing housing shortage (Haughton, 2017; Mace, 2018, p. 23). This was reflected in the BDP Inspector, Roger Clews (2016, p. 44), recognising the 'exceptional, possibly unique' nature of Birmingham's housing shortfall.

More broadly, although each place and region has its own geography and history, meaning that *direct* comparison and 'policy transfer' have limited utility, the West Midlands shares *common issues* with other regions internationally, meaning that there can be important 'lesson learning' (Squires and Heurkens, 2014, p. 2). In many ways, the WMGB is an 'exemplifying case' of the challenges facing regional UGBs (Bryman, 2012, p. 51).

Firstly, it is similar to other non-metropolitan regions, such as the North West (Cheshire's 'golden triangle' and Lancashire's mill towns surrounding Greater Manchester), Yorkshire (the 'golden triangle' and coalfields surrounding Leeds), the Ruhr in Germany and Gothenburg Metropolitan Area in Sweden, in having large disparities between areas with economic deprivation and extensive brownfield land, like the Black Country, and other areas with high levels of economic growth and development pressure on the WMGB, like the M42 Corridor (Figure 2.1; Dorling, 2015; Law, 2000). This enables analysis of the Green Belt's effectiveness in a more varied spatial context than the Greater South East of England, which dominates the literature and policy (Ferm and Raco, 2020), especially in a context with widespread deindustrialisation and brownfield regeneration and, therefore, international relevance. Conversely, the tremendous development pressure in other parts of the WMGB affords scope to explore the challenges of managing housing growth *without* a governance structure like that of Greater London or Manchester (Haughton, 2017).

Secondly, the WMGB is smaller than the MGB, 224,000 ha compared to 514,000 ha, meaning that it is different in spatial character but being a similar size to the West Riding and North West Green Belts (both nearly 250,000 ha), it is comparable to the other Green Belts surrounding England's largest (regional) cities (CPRE and Natural England, 2010, p. 20). Consequently, whilst the project cannot be *statistically* generalised through multiple cases (Scapens, 2004), broader lessons and comparisons can still be drawn from the WMGB (James and Lodge, 2003, pp. 179–190).

Other Qualitative Data Collection Techniques

Interviews

In common with its critical realist approach, this project employed a semi-structured interview approach with similar but slightly different questions at the two spatial scales, national and regional (Winchester, 1999). A common interview schedule reflected the research aims, permitted cross-comparison of responses and was sent in advance to allow participants time for preparation (Saunders *et al*, 2016). However, there were opportunities to add points/comments/observations during the interview whilst the author sometimes varied the interview schedule to go with the 'flow' of conversation and occasionally added supplementary questions (Kvale, 2007).

Sample

Seventy-two interviews were conducted overall with 75 interviewees, including three joint interviews. 63% of the interviews were regional and 37% national, reflecting this project's case study design.

Outside the West Midlands, seven interviews were conducted in the South West of England, two in Yorkshire, one in the East of England, one in the East Midlands and two in the South East.

Using the simplistic split of 'private' to 'other' sectors, 31 interviewees (40%) were from the private sector, whilst the remainder were from other sectors[11]. Again, this was split relatively evenly, although the figures were more complex regarding the non-private sector (Figure 2.2).

The project aimed to interview planners with high(er) levels of seniority as having more experience and an independent voice on the Green Belt, alongside being widely recognised as the 'authoritative' voices of the profession (Slade *et al*, 2019). Nevertheless, the book also wanted to capture the views of young(er) planners, so a young planners focus group was held in Birmingham, and 13% of practising planners interviewed were 'young'[12].

There was not such an even split, with 28% female and 72% male. Although the proportion of male-to-female in the RTPI's membership is more even at 40% female/60% male (Kenny, 2019B, p. 8), this research focused on the Director level, which is less evenly split (Bicquelet-Lock, 2019). However, 86% of attendees at the young planners focus group were female, giving some balance, although gender was not expected to be a key determinant of Green Belt perspectives like one's region (as established in national datasets – see Chapter 7).

Three other participants who were too busy for the interview asked for questions to be answered electronically. Over 100 interview request emails were sent out, but inevitably, not all requests were answered.

Spatial Scale	National 37% (28 interviews with 29 participants)	
	Regional 63% (45 interviews with 47 participants)	West Midlands 46% (33 interviews with 35 participants)
		Other Regions 17% (12 interviews with 12 participants)
Sector	Private 40%	Working for developer/house builder/land promoter 19%
		Working for planning consultancy 20%
		Financial Journalist 1%
	Public etc. 60%	Voluntary sector 12%
		Academia 9%
		Public Sector (Politicians/planners) 40%

Figure 2.2 Composition of interviewees.

Although involving a significant amount of travel, most of the interviews were conducted face-to-face, with the literature generally recommending this as the best way to build rapport with interviewees, gauge their reaction to questions, alongside enable participant observation of office layouts, etc. (Conti and O'Neil, 2007; Harvey, 2011). However, it was impractical to conduct every interview this way, so five interviews were conducted via telephone. Although not as 'natural' as face-to-face, the author and participant 'grew' into the interview and a good rapport developed as Irvine *et al* (2013) found regarding telephone interviews.

Participant Selection

Nationally and regionally, the author focused on the 'triangle' of key stakeholders in the development process: pro-development actors (developer/housebuilder/land promoter), anti-development actors (environmentalist/conservationist) and local/regional/national government (managing these competing interests) (Adams and Tiesdell, 2013; Healey *et al*, 1988, p. 77). Although a useful framework, it is now simplistic given that planning consultancies often work for *both* private sector and public sector clients (Parker *et al*, 2020). Moreover, LPAs themselves may be divided, whereby planning officers are proposing Green Belt release through a local plan, for example, whilst this may be opposed by some local politicians, as in

South Oxfordshire (Mace, 2018). Nonetheless, this multi-scalar triangle of actors can still be identified regionally and nationally, so it formed the main participant identification technique as aiming to explore developer *and* conservationist perspectives. Around this 'triangle' are 'interested parties', so academics, think-tanks and 'expert bodies', such as the TCPA, RTPI and legal profession (Adams and Tiesdell, 2013, p. 60).

Another selection technique was identifying planners and planning stakeholders who had published on, written about or spoken on the Green Belt (such as the Centre for Cities) (Clarke *et al*, 2014). 'Snowballing', whereby planners recommend other planners, colleagues, friends or helpful contacts to interview, was particularly useful as planning is a relatively small profession (Conti and O'Neil, 2007, p. 67; Parker *et al*, 2020, p. 199).

A large number of interviews (72) were conducted because, although 'data saturation' was eventually reached regarding *general* viewpoints on the Green Belt, the later interviews still yielded rich data with a range of geographical insights and a longitudinal aspect with commentary on political developments (Saunders and Townsend, 2016, p. 836).

Focus Group

The focus groups in this study aimed to ask similar questions to the interviews but trigger wider debate and discussion on the Green Belt (Cameron, 2005). Scholars such as Mitchell (1999) and Longhurst (2010) have combined interviews and focus groups as a methodology and way to triangulate (Winchester, 1999). Indeed, as planners often expressed contrasting views on the Green Belt, even from similar private sector backgrounds, the focus group afforded a forum for disagreement and debate (Montell, 2017). Additionally, they explored relevant wider issues, such as land value capture and opposition to development. However, focus groups have been criticised as difficult to manage because certain individuals can dominate the discussion, and other attendees may be reluctant to speak (Cameron, 2005).

Focus Group Process

Two focus groups were conducted: one at a planning consultancy in their office in the South Midlands, focus group I, and another at a global property company in their Birmingham office, focus group II. The aim was to have a mix of planners and surveyors to encourage debate. Focus group I included planners and surveyors (15 in total) with a wide range in ages and seniority, whilst focus group II was seven young planners with a range of roles (from graduate up to principal planner). The focus groups were organised and advertised by 'gatekeepers' internally within the companies, who circulated details of the project summary and questions a few weeks before they took place (Heath *et al*, 2007, p. 403). Both focus groups lasted for over 1 hour 30 minutes, reflecting the enthusiasm and interest of those involved.

Engagement with Campaign Groups

The author contacted regional campaign groups: Project Fields (PF), South Solihull Community Group (SSCG) and Save Stourbridge Green Belt (SSGB). These groups are opposed to the release of land for housing from the WMGB at the rural-urban fringe, so they were poignant cases in exploring campaigners' motivations. It also gave the opportunity for everyday campaigners to give their views rather than just relying on professional campaigners, who were all retired planners. Of course, these groups were not fully representative of the region's geography, but this was largely unavoidable given that campaign groups, and this project was time-limited. The groups represent a range of geographical contexts, from the prosperous northwest (PF (Sutton Coldfield)) and southeast of Birmingham (SSCG (Meriden Gap)) to the industrial Black Country (SSGB (Stourbridge)) (Figure 2.3):

Figure 2.3 Approximate locations in which the campaign groups are operating (adapted from: https://bit.ly/3pQUZdv).

Key: Save Stourbridge Green Belt (SSGB) – opposed to any release of land in the Green Belt for housing in the Black Country Core Strategy Green Belt Review (Dudley withdrew from this plan due to controversy regarding Green Belt allocations).

South Solihull Community Group (SSCG) – opposed to several releases of land in the Green Belt for homes in the Solihull Local Plan (2020) (as this is a Draft Plan, the sites are not displayed in red).

Project Fields (PF) – opposed the release of land in the Green Belt (see red) at Langley for 6,000 new homes.

Project Fields (PF)

PF opposed the BDP's release of land in the Green Belt for housing at Langley and campaigned vigorously (Elkes, 2016). The organisation has remained 'live' and engaged extensively in the subsequent consultation process, so it forms an interesting longitudinal case study of proactive engagement alongside oppositional campaigning, thereby acting in quite an 'agonistic' way (Goode, 2023; Parker and Street, 2015, p. 794). Indeed, PF was affiliated with and supported by the Conservatives in Sutton Coldfield, including the influential MP Andrew Mitchell, and the PF Leader, Suzanne Webb, was subsequently elected as a BCC Councillor and then Conservative MP for Stourbridge in 2019 (until the 2024 Election) covering the area where SSSG is campaigning (Elkes, 2016; Webb, 2020). Webb still frequently campaigns on the Green Belt and as a (until recently) constituency MP, showing the multiscalar character of Green Belt politics (Goode, 2022A; Webb, 2020)[13]. A local politician and PF campaign were interviewed as part of a masters dissertation in 2016, with the politician being re-interviewed for this book and the PF campaigner giving her consent for the material to be reused.

Save Stourbridge Green Belt (SSGB)

This 'live' campaign group was opposed to any release of land for housing from the WMGB within the reviewed and updated Black Country Core Strategy, especially as the Urban Capacity Review (Dudley Metropolitan Borough Council *et al.*, 2018) highlighted that there was not enough 'room' to meet housing need on brownfield land. Consequently, a repeatedly delayed Green Belt Review was undertaken, whereas the 2011 Strategy articulated the policy of a 'strong Green Belt' (Dudley Metropolitan Borough Council *et al*, 2011, p. 42, 2019, p. 35)[14]. However, the plan was very unpopular in Dudley, especially the sites allocated for housing in the Green Belt, so Dudley withdrew from the Core Strategy and is now pursuing its own plan (Goode, 2022B). Although Stourbridge is a relatively 'leafy', middle-class town, the strategic nature of the Core Strategy means that the wider 'industrial' Black Country features prominently in SSGB campaigning. SSGB drew support from the (until recently) West Midlands Metro Mayor, Andy Street, who frequently campaigned on protecting the Green Belt and a brownfield-first policy[15] (Parkes, 2019). Street was a Conservative Mayor in what has traditionally been a Labour Conurbation, winning a small majority in 2017 and 2021 through gaining votes in the wealthy urban-rural fringes (Stourbridge, Solihull and Walsall) (Goode, 2022B).

The author contacted SSGB, and there was a well-written response from 20 respondents who answered in a comprehensive way. These responses were a rich resource for elucidating the views of 'everyday' campaigners.

South Solihull Community Group (SSCG)

SSCG is opposing Green Belt releases in the Draft but repeatedly delayed the Solihull Local Plan (Solihull Metropolitan Borough Council, 2020). One of the professional campaigners interviewed in their home lived near an area of the proposed Green Belt release, which SSCG is campaigning on. The author was given an extensive guided tour around the area, including the affected fields and around the garden, to view the potential development from various angles. Indeed, although an explicitly ethnographic approach was not pursued (Katz, 1994), the author aimed to understand how development in the Green Belt would personally affect campaigners, although it is acknowledged that those living immediately adjacent to developments often have the strongest views (Bradley, 2019A, 2019B).

Other Campaign Groups

A retired Structure Planner, West Midlands and his wife, who was also a planner, were interviewed in their home in the Green Belt. The interviewee drove the author around development on brownfield land in the Green Belt, Hatton Park, to explore how housing developments affect local communities. The Hatton Parish Plan (2013), which includes the views of local residents on the Green Belt, was forwarded to the author to help with the research.

Qualitative Data Analysis

The author listened to recordings of the interviews as part of the data analysis, and the transcripts were read, cross-compared, coded, analysed and searched for key quotes, which formed the backbone of the empirical chapters.

Quantitative Data Collection Techniques

Questionnaire Approach Taken

Questionnaires remain underutilised in planning research, notwithstanding their utility (Goodman *et al*, 2017). The questionnaire developed reflected the author's use of an integrative logic when mixing methods because its aim was to find out as much as possible about the Green Belt from planners rather than capturing a statistically representative sample of their views (Lindsay, 2005; Mason, 2006B). Consequently, getting a representative sample (reliability) and a significant or high enough response rate (validity and generalisability) was not such a *critical* issue, although the author still aimed to maximise the response rate (O'Reilly and Parker, 2013, p. 192).

The themes of the questionnaire's aims were the same as for the other research instruments. However, there was a disappointing response rate of only 9 despite repeated announcements advertising the questionnaire and it being 'open' for over 3 months. This was not a statistically significant sample to carry out data analysis, but the qualitative comments and observations were still helpful, and this attempt at mixed-methods research demonstrates the great challenges of mixing methods successfully and recruiting participants for questionnaires (Lindsay, 2005, p. 121).

Secondary Data Analysis

Secondary data sources formed a key part of data analysis, especially for quantifying and exploring public attitudes towards the Green Belt and housebuilding generally. Some of these questionnaires, such as CPRE's (2015) questionnaire, had the full dataset publicly available so support for the Green Belt could be analysed according to different housing tenures, ages, regions, etc. Alongside spatial and social analysis, similar longitudinal datasets, for example, CPRE's questionnaire in 2005/2015, permitted analysis of temporal attitudinal change. Although there is no space for these datasets to be *fully* presented in the book, they still played a useful role in triangulating and exploring attitudes in a quantitative way (Modell, 2009; Onwuegbuzie and Leech, 2005). There were also questionnaires on general societal attitudes towards housebuilding, like the Social Attitudes Survey (NatCen Social Research, 2017), thus permitting the book to have wider relevance on community opposition towards housebuilding generally.

Wider Material and 'Grey Material'

A large amount of material on the Green Belt was read and analysed, including news articles, policy announcements and updates, the planning press, especially *Planning Portal*, *Planning Resource* and the *Planner*, LinkedIn commentary and blogs by practitioners[16], think tank reports, such as *Rethinking the Planning System* by Policy Exchange (Airey and Doughty, 2020), and local media coverage of Green Belt campaigns, like the Guildford Green Belt Group (Brock, 2018; Munro, 2018, p. 1091; Raco *et al*, 2019, p. 1070). Some of this secondary material was directly included in the empirical chapters (for example, Young, 2020, 2023), but it also formed a key role in challenging and shaping the author's thoughts and findings on the policy (Munro, 2018).

Conclusion

This chapter weaved together the key threads which underpin methodologies: epistemological considerations (critical realism), pragmatic concerns related to the discipline (planning/geography), the topic and the crucial spatial dimension (Bryman, 2012). All three threads are arguably vital in planning research, especially when studying complicated, multi-scalar topics like the Green Belt (Goode, 2022B). More broadly, this study's mixed-methods approach reflects the interdisciplinary nature of planning as drawing in a range of disciplines (Goodman *et al*, 2017; Kilroy, 2017). Moreover, the Green Belt is a particularly rich topic for theoretically conceptualising wider issues related to space and governance whilst being very relevant for practice as probably the most well-known planning policy both in the UK and internationally (Law, 2000). However, a strong, robust methodology, rooted in research aims and drawing on wider epistemological approaches, has been this chapter's main aim (Yeung, 1997). This contextualises and frames the main themes of this book – the Green Belt's history (Chapter 3), its relationship with the housing crisis (Chapter 4), and its conceptualisation (Chapter 5) and the subsequent empirical chapters.

Notes

1 See Peck and Tickell (2003); Jessop (2004) and Peck (2015) for the interconnectedness of different spatial scales and the importance of multi-scalar spatial actors.
2 It is acknowledged that there are other key actors in housing, such as banks and financial investors, but the book primarily focused on the *planning* system (Fernandez *et al*, 2016; Pike *et al*, 2018).
3 Retired planners in the West Midlands have formed a distinctive, professional network, West Midlands Futures, which advocates for a (return) to strategic planning and forms a rich reservoir of knowledge and experience regarding regional planning (Goode, 2022B).
4 Professional' campaigners were often retired RTPI professional planners and 'everyday' campaigners, non-planners, involved in opposing Green Belt development in their area (Amati, 2007; Sims and Bossetti, 2016).
5 The study adopts the government's NUTS definition of the 'West Midlands': the former West Midlands County, Staffordshire Potteries and Herefordshire, Worcestershire, Warwickshire, Staffordshire and Shropshire (Law, 2000).
6 For example, comparing Green Belts in England with those in Wales (and/or) Scotland and Northern Ireland, where Green Belts operate slightly differently (Lloyd and Peel, 2007).
7 Chapter 3 outlines how ribbon/suburban development continued at Barnt Green beyond the city's narrow, inter-war Green Belt in the Lickey Hills (Amati and Yokohari, 2007, p. 317; Self, 1962, p. xii). Birmingham's post-war economic productivity, together with Coventry's, equalled that of the South East, so the government deemed it a 'congested region' and banned office building in 1965 (Hall 2002, p. 88; Hall *et al*, 1973B, p. 520).
8 Chelmsley Wood was ancient woodland, as part of the ancient Forest of Arden, but BCC compulsorily purchased the land, which came into its boundaries, so it was accused of 'annexing' land from Warwickshire (Elson, 1986, pp. 36; Hall *et al*, 1973B, p. 57). New towns were designated at Redditch/Telford (1964), and existing towns, like Droitwich, Stafford and Tamworth, accommodated Birmingham's overspill (Cherry, 1996, p. 152).
9 A Holding Direction is issued by the Secretary of State to prevent an LPA from adopting a local plan to give the Ministry of Housing time to scrutinise it (Boddy and Hickman, 2020, p. 31).
10 The DtC, which was in operation during the period of data collection, has been abolished in the Levelling Up and Regeneration Act, but it is unclear how its replacement, the Alignment Test, will work or what the new Labour Government's policy will be having promised a 'strategic approach' on the Green Belt (Martin, 2024, p. 1).
11 Dividing this figure was difficult, but the planner's sector at the time of interviewing was given to give a contemporary perspective (Slade *et al*, 2019). The non-private sector includes the voluntary sector (i.e. CPRE), institutions (like the RTPI), academia and stakeholders from local/central government, including politicians. The private sector includes developers, planners and land promoters (see Kenny (2019B) for various definitions).
12 Classed as having been a planner less than 10 years from graduation/qualification (Kenny, 2019B).
13 In Webb's (2020, p. 3) maiden speech, she argued, 'When it comes to climate change… I refer specifically to the Green Belt, which is under much pressure in my constituency. I have long championed the protection of the Green Belt, and I know that we can do things differently when it comes to building houses. After all, these greenspaces are the lungs of this great country. If we are serious about climate change, we need to start

thinking differently about how we plan for our future homes and cities and, importantly, how we can protect those vast green lungs with fair funding for remediation and focus on the regeneration of brownfield land'.

14 A joint plan was pursued by the Black Country authorities: Dudley Metropolitan Borough Council, Sandwell Metropolitan Borough Council, Walsall Council and Wolverhampton City Council (Dudley Metropolitan Borough Council *et al.*, 2011, 2019, p. 3). Some of these councils are more urban than others, but as the strategy covered the whole area, it included a Green Belt Review (Dudley Metropolitan Borough Council *et al*, 2019). The Black Country has vast areas of brownfield land, but some of its urban-rural fringe, especially around Stourbridge, is very wealthy.

15 Street argued, 'I simply don't accept this report (the Urban Capacity Review) and will do everything I can to oppose its conclusions' whilst he called Green Belt release the 'easy option' (Parkes, 2019, p. 1).

16 Such as blogs by Mike Best, a recently retired Director at Turley, based in Birmingham, who writes a blog, *Best Laid Plans* (Best, 2019), on the planning situation in the West Midlands, and Phillip Barnes, Planning Director at Barratt Homes.

3   The history of the Green Belt

Introduction

The history of the Green Belt is vitally important for understanding both its temporal longevity and current widespread spatial coverage in England as a relatively distinctive national growth management policy internationally yet one which has experienced policy transfer. This chapter takes a historical institution-alist approach, with Mace (2018, p. 5) helpfully arguing that the Green Belt can be characterised as an 'institution' having 'staying power' as a policy spanning over 70 years, a lobby group in its favour (CPRE), while it is inextricably tied to, and popularly perceived as the guardian of the English countryside. 'Path dependency' has been highlighted by Martin (2010, p. 1) as vitally important in explaining present structures, whilst Valler and Phelps (2018, p. 1) and Sorensen (2015, p. 15) have applied 'historical institutionalism' in planning through ex-ploring how the past 'frames' and constrains future trajectories. Conceptualis-ing the Green Belt policy as an institution and exploring its history through the lens of historical institutionalism not only gives an important historical perspective and context as to its distinctiveness internationally, which is ex-plored through the chapter, but also helpfully highlights possibilities for future policy reform.

The history until World War II is explored in three interrelated sections – a contextual history of urban growth, planning legislation and ideas on urban con-tainment. The Green Belt can be located conceptually in the Industrial Revo-lution (Hall, 1973A), whilst the speed and extent of suburban growth in the inter-war period helped to convert ideas around Green Belts into legislation. The next section explores its history since WWII by weaving together contextual and planning history, whilst the history of ideas surrounding its effectiveness since WWII is discussed in Chapter 4. This chapter's central theme is therefore to briefly chart the growth of British cities and the response to this in the Green Belt to try to manage this growth, which helps to explain the policy's interna-tional distinctiveness.

DOI: 10.4324/9781032674315-3

Historical Context Pre WWII – Urban Growth, Planning Legislation and Ideas Surrounding Containment

A History of Urban 'Sprawl'

A key trend in the history of humanity has been the increasing concentration of population in urban areas such as Babylon, Rome and London (Harvey, 1989, p. 3). However, until the nineteenth century, the size of settlements was *generally* restricted largely by how far people could walk, with towns/cities mainly growing incrementally and slowly in an 'organic', unplanned way in many countries (Whitehand, 2001). The Industrial Revolution rapidly changed the form and function of cities, transforming Britain from a 'prosperous agrarian society' to an 'immensely rich and powerful urban-industrial nation' with its population rapidly increasing from 9 million (m) (1801) to 32.5 m (1901) (Hall *et al*, 1973B, p. 93). As Table 3.1 demonstrates, the Revolution was particularly dramatic in Birmingham and the Black Country, this book's case study, with its rapid relative population increase and the 'heavy' nature of its industries (Goode, 2022A). However, until WWI, the housing growth of London and regional conurbations was generally 'tentacular' or 'ribbon' following key transport routes (Hall, 2002, pp. 23–24; Hall *et al*, 1973B, pp. 21, 83).

In the inter-war era, settlements spread 'outwards' at an unprecedented rate, especially in London and the West Midlands, with 'new' industries developing, such as the motor industry and General Electric (Amati and Taylor, 2010, p. 143). Electric trams, trolleybuses, the Underground and the halt system of railway stations spurred rapid suburban growth and created London's 'circular' urban form through 'filling in' the gaps between key arterial routes (from 1919 to 1939, London's population increased 0.33 times, from 6 to 8 m, yet its urban area increased five times) (Amati and Yokohari, 2007, p. 315; Hall *et al*, 1973B, p. 83). By the 1930s, 148,000 ha of farmland nationally was being used annually for housing; over 100 properties were built each week just in Birmingham and 900,000 new homes were constructed overall from 1921 to 1939 around London (Hall, 2002, p. 27; Manns, 2014, p. 10). This inter-war growth of suburbia alongside the Industrial Revolution earlier in British history generated ideas around the need for a Green Belt, especially for city dwellers to access nature for recreation (Whitehand, 2001).

Table 3.1 Urban population change

	1801	1851	1881	1911	% Growth: 1801–1911
Liverpool	82,000	376,000	611,000	746,000	810%
Manchester	75,000	303,000	393,585	714,000	852%
Birmingham	71,000	233,000	400,774	840,000	1084%
London	1,000,000	2,685,000	4,000,000	6,500,000	550%

From Developed using data from Gracey (1973, p. 78) and Hall (2002, p. 15).

A History of Planning Legislation

Laissez-faire Planning until World War I

Although there were largely unsuccessful attempts to restrict urban growth during the Elizabethan, Stuart and Commonwealth eras, until the 1909 Planning Act, there was little state management of land use, with the Victorian Housing Acts largely focusing on housing conditions (Amati and Taylor, 2010, p. 143; Gracey, 1973, p. 391). Indeed, the 1909 Act focused more on improving and regulating new development through garden suburbs and development plans rather than preventing it outright (Oliveira, 2014, p. 359).

'Loose' Planning in the Inter-War Period

PUBLIC HOUSING

The wartime Prime Minister, David Lloyd-George, was widely believed to have promised 'homes fit for heroes' for soldiers returning from WWI, so low-density social housing was often built in cottage estate style on public transport routes (Hall, 2014, p. 76, 117). These estates regularly accommodated 'overspill' population from conurbations, such as Downham, St. Helier and Becontree (London) and Wythenshawe (Manchester) (Miller, 1992, pp. 183–184). Overall, 763,000 social houses were constructed between 1919 and 1934 (69% of Manchester's total new homes and 47% of Birmingham's, with 90% of Birmingham's social housing built being on greenfield land) (West Midlands Group, 1948, p. 166; Table 3.2). Assuming similar rates of greenfield development by the private sector, this had a tremendous impact on the countryside surrounding Birmingham.

PRIVATE HOUSING

This was also England's largest housebuilding boom, with 1,810,000 homes privately constructed nationally between 1931 and 1939, and the proportion of urbanised land (in England and Wales) increased from 6.7% to 8% from 1918 to 1939 (Thomas, 1963, p. 17). This was partly due to the availability and low price of land and there being multiple, family-run housebuilders. Although there were some high-quality inter-war developments, such as Ealing Garden Suburb and Manor Pool (Harborne), the lack of planning, speed of construction and general poor

Table 3.2 Local authority construction of housing in the inter-war period

Local authority	Number of council houses constructed (1919–1939)	Number of private houses constructed (1919–1939)	Total houses built (1919–1939)
Birmingham	50,268	54,536	104, 804
Manchester	21,979	35,762	57,741

Calculated using data from Hall (2002, p. 20), Hall *et al* (1973B, p. 84) and West Midlands Group (1948, p. 166).

quality architecture deepened popular fears about the countryside being 'swallowed up' leading to a growing conservation movement (Hall, 2014, p. 58). Whilst conservation movements developed in other countries, the scale of inter-war development together with Britain being the first industrial nation, arguably created a particularly powerful conservation movement advocating for measures to restrict urban sprawl.

Attempts to Restrict Urban Growth in the Inter-War Period

Like other past and present governments in liberal democracies around the world, inter-war governments in Britain, which were mostly Conservative, were reluctant to interfere in private property rights, so the only way that a Green Belt could be established was through councils purchasing land to prevent development (Amati and Yokohari, 2003, 2006, p. 126). It was an expensive process, and after the Great Depression, there was neither public money nor political will initially to implement it (Munton, 1983). Nevertheless, Surrey was becoming rapidly urbanised, so, in 1931, Parliament passed the Surrey Local Act giving the council the power to purchase land. In 1932, this power was extended through the Town and Country Planning Act to Middlesex and Essex, although little progress was made, with only 4000 acres of land purchased between 1930 and 1934 (Amati and Yokohari, 2003, 2004, p. 436, 2007, p. 323).

In 1934, Herbert Morrison, a vocal advocate of Green Belts, was elected as the Labour Chairman of the London County Council (LCC), so, in 1935, the LCC Green Belt Scheme began whereby the Treasury, LCC and councils jointly funded land purchase where there was a demonstrable need, i.e. for recreation (Amati and Yokohari, 2003, p. 436, 2004, p. 437, 2007, pp. 317–329). With LCC funding contributing £2 million towards the scheme, 68,000 acres of land around London were purchased by 1939, 19% of Raymond Unwin's proposed 'green girdle', despite huge development pressures (Amati and Yokohari, 2003, p. 6, 2004, p. 434).

Amati and Yokohari (2004, p. 435, 2007, pp. 315–326) have explored the complex, non-linear process by which land was purchased, including civil servants, landowners, campaign groups, 'gifts' of land and secret negotiations to reduce the price of land. For example, Ockham Common, Surrey was suitable land for development but was 'gifted' to the council for only £24,000 by its owner, Lady Lovelace (Amati and Yokohari, 2007, pp. 336–337). The 337-acre Lambourne Hall, Essex, was too expensive for the council to purchase at £105/acre, so it was sterilised from development for £54/acre (Amati and Yokohari, 2007, p. 322). In Birmingham, the council purchased the land owned by the Cadbury family in the Lickey Hills (West Midlands Group, 1948, p. 211).

In 1938, the Green Belt Act was passed, which protected the land purchased since 1935 (Lloyd and Peel, 2007, p. 645; Shaw, 2007, p. 576). Although much smaller than the current Metropolitan Green Belt and not forming a continuous 'ring', for the first time, London had a Green Belt and, together with the Ribbon Development Act (1935), urban growth was more effectively restricted (Hall, 1973A, p. 386; Hall *et al*, 1973B, p. 106). Council ownership of the land also meant that it could be used more effectively for recreation than the largely privately owned land of the post-war Green Belt[1] (Sturzaker and Mell, 2016, p. 35). Indeed, notwithstanding the limitations of the inter-war Green Belt, Inch and Shepherd (2019) have underlined the central importance of 'conjunctures' in planning history, and the unrestricted growth of the inter-war can be arguably seen as a key reference point or 'conjuncture' in the subsequent and continuing justification for the post-war Green Belt in England.

A History of Ideas: Key Thinkers and Ideas Surrounding Urban Containment

Although Amati and Yokohari (2007, p. 311) have warned that the policy's history has been too focused on 'planning heroes', the development of ideas is still invariably associated with key thinkers (Hall, 2014). This section looks mainly at Ebenezer Howard, Raymond Unwin, and Sir Patrick Abercrombie, although the Green Belt's current context and form/function is different from what they all envisaged.

Garden Cities of Tomorrow

The Green Belt can be located conceptually in the Industrial Revolution when thinkers started to give serious consideration to restraining urban growth with the countryside being romanticised by poets, such as Thomas Hardy and the Arts and Crafts Movement, alongside popular concern regarding urbanisation and the poor quality of urban life (Hall, 2014, p. 190; Miller, 1992, p. 189). In 1905, Henrietta Barnett complained that London was extending its 'long and generally unlovely arms' into the countryside (Hall *et al*, 1973B, p. 82). Birmingham was described, in 1902, as 'continuous roads and houses from Aston on the east to Wolverhampton in the west...quite as much entitled to a single name as is Greater London' although it was generally viewed as better planned than London (Hall, 2014, p. 96). These factors influenced Howard, who, in his famous *Garden Cities of Tomorrow* (1902) (Howard, 1946, pp. 44), advocated for garden cities:

> To restore the people to...that beautiful land of ours...the very embodiment of Divine love to man...on its bosom, we rest.

Hall (2014, p. 98) argued that Howard's genius was exemplified in the *Three Magnets Diagram,* which conceptually brought the rural and urban *together*. Indeed, Howard (1946, p. 47, 49) was troubled by the 'unholy separation of society and nature...Town and country must be married', but he differed from contemporaries in recognising the benefits of cities, including 'high wages...tempting prospects of advancement...places of amusement'. Consequently, Howard did not advocate preventing urban growth *per se* but that it should be planned in a 'cellular' way through self-sufficient garden cities (Howard, 2003, p. 8, 191). The settlement would be surrounded by a Green Belt or 'outer ring', a country/agricultural 'belt' owned by the Garden City Corporation for 'large farms, small holdings, cow pasture' and people's recreational use (Howard, 1946, p. 55).

Howard's *Social Cities* diagram in *Tomorrow: A Peaceful Path to Real Reform* (1898), which advocated the development of multiple garden cities to accommodate population growth, was not printed in *Garden Cities of Tomorrow* (1902). In a 'subtle yet profound transmogrification', Howard was misinterpreted by Abercrombie as advocating garden cities and Green Belts as an ideal '*end-state*' or 'fixed blueprint for the future' rather than a *dynamic, evolving process* (Hall, 1974, p. 42, 2014, p. 98; Keeble, 1971, p. 70). Consequently, Howard's ideas regarding Green Belts 'froze' and have proven 'incredibly resistant to change' as still revered as the 'first article of the planning creed' around the world, despite being developed in a different era to even the Abercrombie plan, with the low price of land, fears of rural depopulation and rapid future urban economic and population growth excepted in Howard's time (Gant *et al*, 2011, p. 266; Hall *et al*, 1973B, pp. 46, 107). Nonetheless, Hall *et al* (1973A, p. 71) contended that it is 'misleading to call Howard's ideas a Green Belt at all' because Howard proposed 'cluster development' rather than Unwin's/Abercrombie's *continuous* Green Belt around *existing* conurbations.

Other Key Thinkers: Ashbee, Pepler and Aston Webb

There were other influential individuals, contemporary to Howard, involved in the London Society, which was a ferment of intellectual planning ideas (Table 3.3) (Hall, 2014, p. 190).

Table 3.3 Other key thinkers/documents

C. Ashbee (Architect)	In 1894, the London Survey Committee was established, taking inspiration for Green Belts from the parkways of the City Beautiful Movement and Vienna's Ringstrasse. Influential to Howard and Unwin who wrote, 'in America and on the Continent…trams run along a belt of grass'.
Lord Meath, Head of the LCC; Sir George Pepler	Meath was inspired by the avenues of Boston/Chicago, and Pepler called for a ¼ mile-wide 'green girdle' around London.
William Bull, Head of LCC's *Parks and Open Spaces Committee*	Called for a 'circle of green sward and trees' ½ mile wide around London for transport, as in Europe, but Bull still proposed development beyond this Green Belt.
David Niven, Architect	In 1910, he called in the *Architectural Review* for a 'continuous garden city around London that would be a healthful zone'.
Aston Webb, Architect	Addressing the London Society in 1914, he claimed to have dreamt that there would be a 'belt of green all-round London' of 'open spaces' in 2014.
Development Plan of Greater London (1919)	Called for 'belts of green parkways' around London.

From Derbyshire (2015, p. 2), Manns (2014, pp. 6–10), Oliveira (2014, pp. 360–362) and Whitehand (2001, p. 49).

Unwin and 'Green Girdles'

A key London Society figure and follower of Howard was Raymond Unwin, who was the main advocate for Green Belts in the inter-war period (Miller, 1992, p. 200). He argued in a book with Niven, *London of the Future*, for a:

> Continuous Green Belt completely encircling London proper…To protect its inhabitants from disease, by providing fresh air, fresh fruit and vegetables, space for recreation and contact with…nature…such areas should be as quickly as possible reserved, and to a generous extent, to form a green belt.
> (Cited in Derbyshire, 2015, p. 7)

The inter-war growth of London spurred further popular resistance, exemplified in Clough Williams-Ellis's *England and the Octopus* (1928), with Unwin continuing to make his case for Green Belts. In 1929, the London Planning Committee was established to examine Unwin's recommendation for an 'agricultural belt' of 6 miles wide (Thomas, 1963, p. 16). Unwin argued for:

> A green belt or girdle…a background of open country…as near to the completely urbanised area of London as practical…[to]…provide a reserve supply of playing fields and other recreation areas and of public open spaces.
> (Cited in Manns, 2014, p. 12)

Nevertheless, despite the Second Report of the Committee being published (1933), its findings were largely unheeded, although a form of Green Belt had been established before Unwin's death in 1942 (Amati and Yokohari, 2007, p. 316). Indeed, Abercrombie (1944, p. 11), in the *Greater London Plan*, wrote that it 'unhesitatingly adopted' Unwin's arguments for a 'continuous green background of open country'.

Green Belts and New Towns Triumphant

Abercrombie translated Howard's and Unwin's ideas on Green Belts into a concrete plan. In 1941, he was commissioned to plan for the post-war reconstruction of London and stipulated that he would also plan for the Greater South East (Wannop and Cherry, 1994, pp. 35–38; Whitehand, 2001, p. 49).

There was consensus, especially after the Blitz, that the decentralisation of the cities was required alongside new towns and Green Belts. Hall *et al* (1973A, pp. 55–56) located Abercrombie in the genteel, Oxbridge-educated circle governing post-war Britain, who saw change as slow and adopted the paternalistic ethos of 'experts' knowing best. Consequently, Abercrombie's plans and the 1947 Act largely did not foresee the population/economic growth and mass car ownership of the post-war era, with change viewed as controllable and manageable through planning and Green Belts (Hall, 2002, p. 108). Abercrombie (1944, p. 30) argued for 'decentralisation not growth' and that 'it is better to err on the side of being too restrictive', with Table 3.4 setting out the ideas in his plans.

Table 3.4 Abercrombie plans

Plan	Scope	Main ideas regarding the Green Belt
County of London Plan 1943	Transport focused and proposed a series of orbital railways and roads.	'*To preserve a broad area of unspoilt country within easy access for London's inhabitants... primarily for agriculture with no further building other than that ancillary for farming... the Green Belt Ring, with its open lands and running streams used for recreative purposes, and acting as a barrier to the continuous expansion of London... A stretch of open country at the immediate edge of the unwieldy mass of building is imperative'*. [Towns in the Green Belt] '*will be greatly enhanced by their permanent setting in open surroundings, free from the menace of coalescence with one another in a drab sea of building*' (Forshaw and Abercrombie, 1943, p. 24).
Greater London Plan 1944	Focused on urban form. Proposed four key areas: **Inner Urban** (inner-city London) **Suburban** (inter-war London) **Green Belt** (no building) **Outer Country** (location of new towns)	'*A gigantic Green Belt round built-up London... primarily for the recreation* [and]*...fresh food, and to prevent further continuous suburban outward growth... a careful line has been drawn round the largely built up sections... The most important need is the linking up of open spaces... there should be a pedestrian system of communications as efficient as that for the motor... the* [Green Belt] *should be controlled in such a way that landscaping, afforestation... full public access use may be harmonised*' (Abercrombie, 1944, pp. 11–35, 109–111).

The key thread running through Abercrombie's plans was containing London's growth through a 5–7 mile wide Green Belt (Figure 3.1) (Hall *et al*, 1973B, p. 51). It would have a 'dual use' for agriculture *and* recreation whilst the *Greater London Plan* made provision for 125,000 people to be decentralised to peripheral, satellite LCC estates, such as Harold Hill, alongside 'very limited expansion of those towns already in it' (Abercrombie, 1944, pp. 34–35, 111).

The centrepiece of Abercrombie's plans (1944, p. 78) was for 1–1.25 million people to be decentralised from London to a 'necklace' of eight new towns (Wannop and Cherry, 1994, p. 38; Hall, 2014, p. 216). Abercrombie's Plans were so influential that they remain the basis for the current London Plan, with its Green Belt protection, albeit that now, due to the rising population, there is a greater focus on urban densification (Lainton, 2014; Mace, 2018; Manns, 2014). Indeed, Abercrombie travelled widely, and, in an interesting example of policy transfer, his ideas were extremely influential throughout the Commonwealth and the rest of the world, as seen, for example, in the Green Belt introduced in Tokyo and Bangkok (Amati, 2008; Yokohari *et al*, 2000).

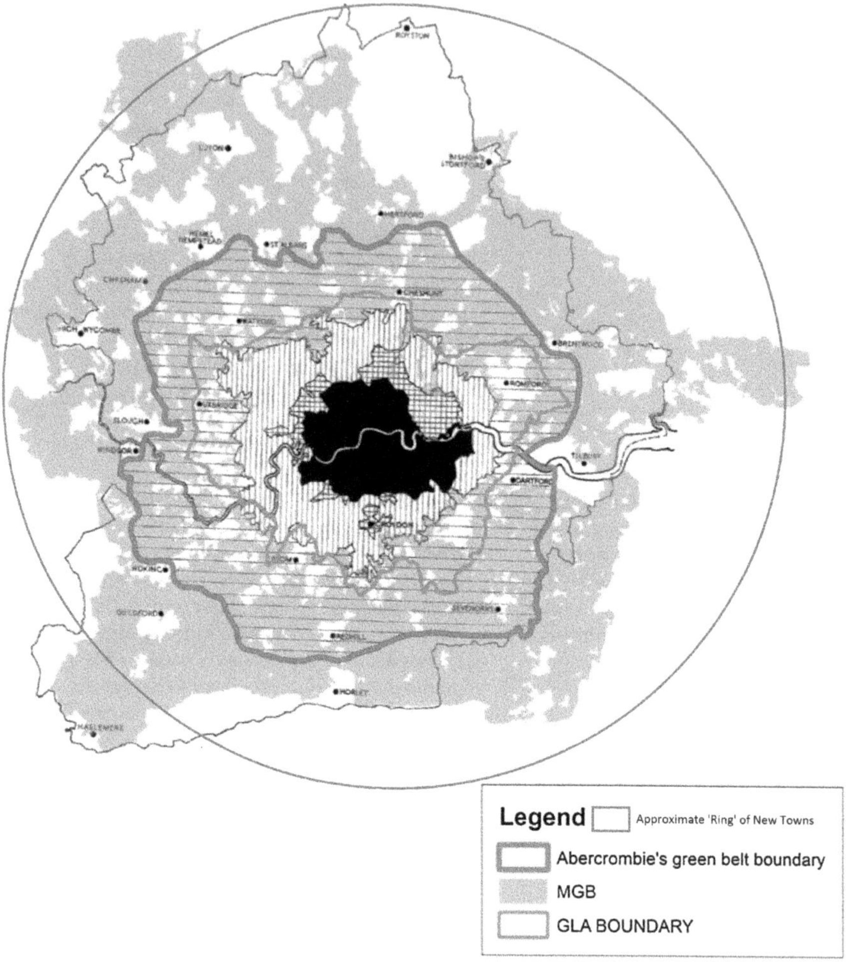

Legend — Approximate 'Ring' of New Towns

Abercrombie's green belt boundary

MGB

GLA BOUNDARY

Figure 3.1 Abercrombie's Green Belt compared to today's Green Belt (Adapted from Mace, 2018, p. 8).

The Green Belt's Establishment and Ascendancy

Background to 1947 (Town and Country) Planning Act

These key Reports had an important impact on the Green Belt and 1947 Act, so they are summarised in Table 3.5.

1947   Town and Country Planning Act

The 1947 Act created the modern planning system with development rights being nationalised and a statutory requirement for districts to prepare plans allocating

Table 3.5 The Barlow Commission, Scott Report and the Uthwatt Committee

Report	Findings/Recommendations	Impact upon Green Belt
Barlow (Royal) Commission (*On the Distribution of Industrial Population*) (1940)	Regional disparities and the 'gap' between London and 'distressed' regions needed addressing through controlling industry/housebuilding.	Took a negative attitude towards growth despite quality of life improving in urban areas with Abercrombie, as a Committee member, arguing for controls. Growth had: '*despoiled the countryside and largely diminished* [its] *health-giving elements… the most insidious menace*' (HMSO, 1940, p. 226).
Scott Report on Land Utilisation in Rural Areas (1942)	Following the threat of starvation in WWII, the best *and* medium quality agricultural land and rural way of life should be preserved.	Agricultural land should be preserved, so the Green Belt should be primarily for agriculture, not recreation (Abercrombie used an Agricultural Classification system in his Plans). Professor Dennison highlighted that agriculture had become more productive but the Report was influenced by the Geographer Dudley Stamp who referred to land's '*chief characteristic*' being its '*attractive patchwork appearance*' so 'openness' was included in the Green Belt policy (MWP, 1942, p. 3). The Report proposed a '*tract of ordinary country, of varying width, round a town*', forming the key inspiration for Abercrombie's Green Belt (MWP, 1942, p. 72).
Uthwatt Committee on Betterment (1942)	Government should purchase/nationalise development rights on undeveloped land.	The state was expected to undertake most of the development. However, the Green Belt remained in private ownership and land value capture continues to be a thorny issue (Cheshire, 2009).

Hall (1973A, p. 49, 2002, pp. 105–106), Ministry of Works and Planning (MWP) (1942) and Thomas (1963, p. 79).

land for development and to update them every 5 years (Cherry, 1996, p. 148; Hall and Gracey, 1973, p. 99). This was different from other countries, like Japan, where the post-war Constitution, influenced by the USA, enshrined the right for private landowners to develop land (Amati, 2008).

Short *et al* (1987, p. 31) argued that the planning system was established to resist urban growth and to protect the countryside and rural way of life through the Green Belt, whilst Hall *et al* (1973B, p. 40) argued that an 'unholy alliance' formed between urban authorities, wanting to protect their tax base and support urban redevelopment, and rural shires wishing to maintain their rural exclusivity. The failure to also reform the county system of governance and local government boundaries intensified subsequent political struggles over the Green Belt, with Green Belt boundaries often being drawn tightly around Labour cities surrounded by Conservative counties (Dockerill and Sturzaker, 2019; see also Chapter 9).

The Philosophical and Political Context of the Green Belt

Much of the long-standing, popular antipathy to urban growth in England arguably stems from the dominance of the gentry in British society, leading to the countryside being idealised and this percolating down to fashion a society in which homeownership, especially in semi-rural locations, is idealised as owning one's own 'castle' or 'estate' (Bunce, 1994; Hall, 2014; Matless, 1998, p. 365). Moreover, Britain being an island with a relative and perceived shortage of land and the world's first industrial nation explains the rural shires subsequent resistance to housing growth, such as Cheshire, Worcestershire and Warwickshire, with fears of being 'threatened' and 'spoiled' by the encroaching industrial cities, like Birmingham and Manchester (Elson, 1986, p. xiii, 40; Goode, 2022B). This explains differing attitudes towards development in countries such as Scotland, Ireland, France and the USA, which are popularly perceived to have more 'space' (Amati, 2007; Lloyd and Peel, 2007).

Prior and Raemaekers (2007, p. 580) therefore conceptualised the distinctiveness of the Green Belt in England as a product of the post-war, Fordist economic system in which the state played a large part in managing and planning the economy, as evident in the amount of state intervention in the 1947 Act and Green Belt policy compared with the 'loose' inter-war planning system (Dorling, 2015; Martin, 2010). The Green Belt was also one of the 'triptych' of post-war planning policies alongside new towns and regional policy (Mace, 2018, p. 10). However, the fact that the Green Belt has remained a constant and popular policy, despite successive waves of neoliberalism rolling up other parts of the 'triptych', is evidence of the enduring strength of the conservation movement in England (Scott, 2015; Sturzaker and Mell, 2016).

Green Belts Codified and Implemented – 1955 Circular 42/55

The Green Belt was codified and its purpose outlined by the Housing Minister, Duncan Sandys, in 1955 (Elson *et al*, 1994, p. 154). Sandys, as a genteel Conservative, was enthusiastic about encouraging councils to establish them although, rather than being primarily for agricultural or recreational purposes, the Circular stated their purpose as *'preventing urban sprawl'* (Amati and Yokohari, 2007, p. 315; Hall *et al*, 1973A, pp. 52–57):

- [To] check the further growth of a large built-up area;
- Prevent neighbouring towns from merging into one another; or
- Preserve the special character of a town.

(Ministry of Housing and Local Government, 1955, p. 1).

Circular (42/55) was only substantially altered in 1984 when urban regeneration was added as a purpose whilst preventing urban sprawl still forms the policy's central justification (Sturzaker and Mell, 2016, p. 31). Circular (42/55) was reinforced by Sandys' successor as Housing Minister, Henry Brooke, who argued in 1961 that:

> The very essence of a Green Belt is that it is a stopper. It may not be very beautiful...but without it the town would never stop.
>
> (Cited in Hall *et al,* 1973B, p. 58)

In 1946, the government accepted Abercrombie's plan and, in 1950, produced a Green Belt map based on it (Cherry, 1996; Munton, 1983, p. 147). This was implemented between 1954 and 1958 as the Home Counties incorporated this Green Belt into their plans (Whitehand, 2001, p. 50). However, Circular (42/55) encouraged the Home Counties to designate *further* land as Green Belt and the regions to introduce Green Belts to protect the countryside with another Circular (50/57) permitting this (Longley *et al*, 1992, p. 438; Ministry of Housing and Local Government (MHLG), 1957, p. 1). The Green Belt quickly became so popular that they gained a 'life of their own' in terms of expansion, although Sandys himself believed they should be only 5–10 miles wide (Manns, 2014, p. 14). Elson (1986, p. 19) examined 69 proposed Green Belt sketches in local plans submitted between 1955 and 1960 and found that the Green Belt's primary objective was to be a 'stopper'. Consequently, from 1968 to 1984, the amount of land in the Green Belt increased from 693,000 to 1,581,500 ha or 12% of England's land surface (Sturzaker and Mell, 2016, p. 36)[2].

The urgency for the post-war expansion of the Green Belt by councils arose from fears that the government would establish new towns in their areas and the large increase in housebuilding in the 1960s/70s (*Ibid*, p. 35). The Labour Government (1964–1970), while still defending the Green Belt, was reluctant to add more land to it with its Housing Minister, Richard Crossman (1964–66), concerned about projected population growth, so he would not approve development plans until regional studies on land availability were completed (Hall *et al*, 1973A, p. 57). Consequently, the Green Belt only expanded 184,000 ha between 1964 and 1970 (Elson, 1986, p. 39).

A turning point was Edward Heath's Conservative Government (1970–74) when Peter Walker was Environment Secretary (1970–1972) (Elson, 1986, p. 44). No more new towns were designated, apart from Stonehouse, development pressure on the Green Belt reduced as population forecasts were lowered and the focus shifted towards inner-city problems (Cherry, 1996, p. 148). Furthermore, Walker revived the policy of designating new areas as Green Belt so maps were 'tightened up' and settlements 'washed over' to prevent further development, with the government largely maintaining the county system of governance in the 1972 Local Government Act (Sturzaker and Mell, 2016, p. 34). In 1971, the government approved large Green Belt extensions covering most of Surrey, increasing the overall original Green Belt by 20%, and in Kent and Buckinghamshire, increasing it by 15%, in what Elson (1986, p. 40) characterised as a 'victory for the Homes Counties'.

Green Belts in Practice (1950s–70s) – Tension between Town and Country

However, although the *general* trend was to uphold the Green Belt policy, in the 1960s/70s there was also often a pragmatic attitude towards it and notable compromises (Gant *et al*, 2011, p. 266). The government allowed significant transport development in the Green Belt, including Britain's motorway network, like the M25 and M42, whilst airports, such as Manchester, Birmingham and Heathrow, have expanded. Additionally, growth has occurred around them, such as the World Logistics Park (Manchester) and National Exhibition Centre, alongside adjacent business parks, like Blyth Valley and Birmingham Business Park (Law, 2000, p. 64; Kells *et al*, 2007, p. 7).

A significant area of contention was where the overspill population from the large cities should be located, with new towns and existing towns expanded under the 1952 Town Development Act being insufficient to meet housing needs (Sturzaker and Mell, 2016, p. 28). Crossman therefore approved developments in the Green Belt, like Chelmsley Wood (Birmingham) (1964) (as outlined in Chapter 2), and private sector new towns, such as New Ash Green (Kent) (1966) (Ward, 2005B, p. 346).

Thatcher's Government (1979–1990) – Attempts at Reform and Retreat

Despite the Thatcher Government's rhetoric around 'rolling back the frontiers of the State' and *Lifting the Burden* (the title of a White Paper (1985)), the amount of land in the Green Belt increased by 45% in Thatcher's first term, more than doubled from 7000 km^2 (1974) to 16,000 km^2 (1984) (Ward, 2005B, p. 336). The total amount of land involved in the Green Belt's expansion in the 1980s/90s was seven times more than that used for new housebuilding (Abbott, 2013, p. 18).

Ward (2004, p. 226) has called the Green Belt the 'exception to Thatcherism' although there was constant tension within the Conservative Party, which endures to this day, between the neoliberal, pro-development lobby and the conservationist, rural lobby (Tait and Inch, 2016, p. 187; Ward, 2005B, p. 329). For example, in 1983, the Environment Secretary, Patrick Jenkin, produced a Departmental Circular on the Green Belt and land for housing, which suggested modest reforms for more flexibility regarding Green Belt 'swaps', i.e. adding more land in exchange for land being taken out of the Green Belt (Munton, 1986, p. 211). However, such was the political opposition from the environmental, countryside lobby, especially CPRE, and backbench Conservative MPs, that Jenkins retreated and issued a White Paper in 1984 (Circular 14/84) which stressed the primacy of the Green Belt (Gant *et al*, 2011, p. 267). Even though the completion of the M25 in 1986 opened up huge development opportunities, planning inquires upheld Green Belt protection (Gallent and Shaw, 2007, p. 620; Munton *et al*, 1988, p. 327).

In 1983, a group of housebuilders formed Consortium Developments Ltd., which hoped to build 15 private sector new towns, each around 5000 dwellings, having been emboldened by the Thatcher Government's anti-planning rhetoric[3] (Ward, 2005B, p. 342). One of these, Tillingham Hall, Essex, was proposed in 1985 in the

Metropolitan Green Belt (Munton, 1986, p. 212). Nicholas Ridley, the Environment Secretary, was one of Thatcher's most free-market Cabinet Ministers, so it was widely expected to be approved[4] (Cherry, 1996, p. 201). However, there was widespread opposition at the public inquiry, which Hall (2014, p. 442) argued became one of the '*causes célèbres*' of planning history. In 1986, Ridley promised that the Green Belt is 'safe in our hands' and 'sacrosanct' so Tillingham Hall was rejected demonstrating the political nature of the policy (Blake and Golland, 2003, p. 143).

Ward (2005B, p. 394) helpfully highlighted that opposition to Consortium Developments was particularly fierce because of the aggressive way it acted in seeking to override Local Planning Authorities (LPAs). Although still contentious, where large developments proposed in Green Belts had LPA support, such as Bradley Stoke (8500 homes, South Gloucestershire), Chafford Hundred (5000 homes, Essex) and Church Langley (3500 homes, Harlow), these were approved by Ridley in 1986/7 (Munton, 1986, p. 212; Ward, 2005B, p. 352). Nevertheless, Consortium Developments expected the Thatcher Government to override community and LPA opposition as it used central power to overrule trade unions, Labour councils and inner-city areas, but the government did not have the 'heart' to do this to its 'own side' – Conservative constituencies, councils and communities (Hall, 2014, p. 442; Ward, 2005B, p. 354).

Major's Government (1990–1997) – Green Belts Triumphant

The fierce planning battles and urgency of housebuilders to build on Green Belt (and greenfield) sites somewhat receded in the 1990s due to the recession, lack of house price growth and the (relatively) abundant supply of 'easy' brownfield sites for housing near town/city centres (Breheny, 1997; Law, 2000, p. 60). In 1995, the government set a target of 50% of development on brownfield land, which was raised to 60% before the 1997 election (Breheny, 1997, p. 209). This stemmed from compact cities and sustainable (brownfield) development becoming the dominant policy rhetoric and objective (PPG13), especially with the increasing importance of environmental issues following the Rio Earth Summit (1992) and the White Paper, *This Common Inheritance* (1990) (Warren and Clifford, 2005, p. 355). Ward (2005B, p. 354) argued that it gave people who wished to see rural housebuilding restricted legitimacy, whilst Pennington (2000, p. 73) described the increasingly plan-led and protectionist nature of planning during the 1990s as a 'victory' for the increasingly influential environmental movement. Moreover, development in the Green Belt was further restricted through PPG2 (1995, p. 5), which stressed 'safeguarding the countryside from (further) encroachment' as a key aim (Gant *et al*, 2011, p. 267). Consequently, between 1989 and 1997, the Green Belt expanded another 1038 km^2 (Sturzaker and Mell, 2016, p. 40).

The Green Belt and the Housing Crisis (1997–2024)

This period has been characterised by an increasing awareness of the worsening housing crisis, leading to numerous attempts at 'solving' this crisis, largely through planning reform. Some of these have affected the Green Belt, but there has been no direct, fundamental 'attack' on the policy by the national government yet (Amati, 2007, 2008; Sturzaker and Mell, 2016, p. 42).

New Labour Government (1997–2010)

Part of New Labour's electoral success was based on promising to protect the Green Belt and appealing to middle-class Home County voters (Tallon, 2013). It also wanted to appeal to urban voters, so the Urban Task Force was established in 1999, calling for an 'urban renaissance' through the revival of cities to house 4 million more people and a 60% brownfield land housebuilding target (Allmendinger, 2009; Nathan, 2007, p. 5; Sturzaker and Shucksmith, 2011, p. 176). Nevertheless, with growing awareness of the housing crisis and the government's focus shifting towards social equity and sustainable development, it committed to building more homes in the Greater South-East, including the Thames Gateway, and beyond the Green Belt in places such as Ashford, Milton Keynes and Northampton (the 'South Midlands') (Amati, 2008, p. 9; Gunn, 2007, p. 607; Shaw, 2007, p. 576).

The Treasury commissioned the Barker Review (2004, p. 121), which highlighted the need for 200–250,000 new homes annually to keep prices stable. Barker (2004, p. 44) argued that greater flexibility was needed regarding the Green Belt, including the need to review boundaries regularly and explore the possibility for 'swaps', although she did not call for its abolition or significant reform. Barker (2006, pp. 9–10) also highlighted that the Green Belt was often 'low value agricultural land/landscape quality [with] limited public access' so a 'more positive approach' was required to enhance it. The government was reported to have seriously considered relaxing the designation after the Barker Review, and, although significant relaxation was deemed too controversial, a more flexible approach was adopted with plans de-designating land in the Green Belt in exchange for adding land being approved in the Homes Counties, like Bedfordshire, in 2008 (Gallent and Shaw, 2007, p. 618; Gunn, 2007). The Green Belt Direction (Circular 11/05) granted powers to John Prescott, the Deputy Prime Minister, to call in Green Belt developments, thereby giving central government greater scope to override local opposition (Gant *et al*, 2011, p. 261). However, the Regional Spatial Strategies generally upheld the Green Belt in post-industrial regions, like the West Midlands, although, in the South-East Plan, there were controversial proposals, like a 4000 home extension to Oxford (Gant *et al*, 2011, p. 267).

The Coalition, Conservative and Labour Government – 2010–2024

The Conservatives' electoral campaign in 2010 partly rested on their campaign against Regional Spatial Strategies (RSSs), housebuilding generally and Labour's supposed 'attack' on the Green Belt (Lowndes and Pratchett, 2012; Sturzaker, 2011, 2017, p. 26). Conservative-led governments since 2010 arguably sent mixed messages about housing reflecting the key, competing strands in Conservative thinking – anti-development, one-nation localism and pro-development (neo)liberalism (Allmendinger and Haughton, 2011, p. 317; Tait and Inch, 2016, p. 187).

The *rhetoric* of localism, neighbourhood planning, abolishing housing 'targets' and regional planning (RSSs in 2010/11) fostered a more anxious, active and protectionist culture, especially regarding Green Belt protection, with a large 'void' existing in strategic planning (see Chapter 9) (Sturzaker, 2011, p. 555).

This was encouraged by political promises to protect the Green Belt, such as the former Prime Minister David Cameron constantly promising that the Green Belt was 'safe in our hands'. For example, the more frequent refusals of appeals by Secretaries of State in the run-up to elections, like the Communities Secretary, Eric Pickles, refusing notable applications in the Green Belt, such as Coventry Gateway (2015) (Edgar, 2015, p. 1). Likewise, in 2019, former Prime Minister Boris Johnson backed the Wolverhampton Green Belt campaign, whilst protecting the Green Belt was one of the few areas where Conservative and Labour Manifestos agreed in the 2019 General Election[5] (Brassington, 2019).

However, with the deepening housing crisis, the government initially pursued a more 'muscular localist' approach as focusing increasingly on housebuilding with the (then) Chancellor, Philip Hammond, setting an annual 'target' of 300,000 new homes in 2017, which remained in the 2019 Manifesto (Conservative Party, 2019, p. 31; Tait and Inch, 2016, p. 187). 252,500 new homes were completed in 2022 from a historically low figure of 122,000 completions in 2021 with the industry slowly recovering from the financial crisis and output hit in 2020/21 by Coronavirus and the lockdowns. However, completions appear to have stagnated recently with the cost of living crisis and a reduction of planning permissions granted (Buckle *et al*, 2024, p. 1).

Until 2021, the government's efforts to solve the housing crisis largely revolved around increasing housing demand through policies such as 'Help to Buy' and First Homes, planning reform with around 200 reforms since 2010 and the pro-growth National Planning Policy Framework (NPPF) (Department for Communities and Local Government (DCLG), 2012; The Economist, 2017, p. 1). It also dismissed key reports or campaigns on reforming the Green Belt, including Urbed's (2014) Wolfson Prize winning scheme for garden villages[6], the Labour MP Siobhain McDonagh's campaign to allow development in the Green Belt around railway stations and a report on reclassifying and reviewing the Green Belt (2019), *Raising the Roof*, by the influential (former) Conservative MP Jacob Rees-Mogg[7] (Airey and Doughty, 2020; CPRE, 2018).

In 2020, the government published a Planning White Paper, *Planning for the Future*, which proposed moving to a more rules-based, less discretionary system, although the government still proposed maintaining existing Green Belt protections (Airey and Doughty, 2020; Simons, 2020). At the same time, it published changes the way that housing numbers are calculated, the Standard Method, which would potentially have put more development pressure on largely Conservative voting constituencies in the South East where house prices are higher (Young, 2020). This began a significant political backlash against planning reform, including the 2021 Amersham and Chesham by-election, whereby the Conservative Party's large majority was overturned largely due to opposition to national housing targets and planning reform in a Green Belt area (Yorke and Neilan, 2021).

In response, both the government's rhetoric and policy stance towards planning evolved with, firstly, the Planning White Paper being shelved and the former Prime Minister Boris Johnson making comments in 2021 about not building on greenfield sites in response to electoral challenges in the 'blue wall' of Conservative constitencies in the Home Counties (Riley-Smith *et al*, 2021). Secondly, the previous government moved housing targets to being voluntary in 2023 and tilted the NPPF to being less pro-growth as making Green Belt review voluntary in local plans (DLUHC, 2023).

In the lead up to the 2024 election, the Labour Party became increasingly pro-growth and anti-planning, with Sir Keir Starmer arguing for building on the so-called 'grey belt' in the Green Belt, whilst the (formerly) governing Conservative Party promised to protect the Green Belt (Martin, 2024; Mason, 2023). The policy of the Labour Government has been set out in a revised NPPF (published December 2024) which confirms proposals allowing housing development on land designated as 'grey belt' alongside a significant uplift in housing targets in many local authorities with Green Belt through the Standard Method (Ministry of Housing, Communities and Local Government (MHCLG), 2024). Whilst it remains to be seen how this new NPPF operates in practice, this could led to significantly more housing development in the Green Belt.

However, going forward *wholesale* Green Belt reform is unlikely because the Conservatives have generally been supportive of the Green Belt, Labour's proposals about building on the 'grey belt' are arguably modest and widespread popular and political opposition has historically hindered previous attempts by Labour governments at significant Green Belt policy reform (Inch *et al*, 2020; Sturzaker and Mell, 2016)[8]. This demonstrates that, whilst there is growing political recognition of the housing crisis in both of the main political parties, the preferred policy 'solution' is planning deregulation whilst upholding the Green Belt in general terms (Buckle *et al*, 2024; Goode, 2022B; Mace, 2017; Sturzaker, 2017).

Conclusions from the History of the Green Belt

This chapter has explored the widespread, popular and idyllic image of the English countryside, of which the Green Belt is seen as a guarantor and bastion and an arguably particularly English antipathy towards urban 'sprawl', probably deriving from the magnitude and rapidity of the Industrial Revolution (Lloyd and Peel, 2007; Matless, 1998, p. 641). Amati and Yokohari (2007, p. 312) highlighted how the Green Belt has used 'myths about the past to justify the present'. This is important because the Green Belt remains a very influential policy internationally, yet attempts have been and are being made to introduce similar policies, like Urban Growth Boundaries, in very different institutional contexts such as Southeast Asia, Oregon and Medellin (Amati, 2007; Boyle and Mohamed, 2007). Nevertheless, there are important lessons which can be learnt from the English experience in terms of theorisation, including the importance of historical institutionalism, alongside highlighting shared characteristics with similar constructs such as the Groene Hart (Green Heart) in the Netherlands and the Golden Horseshoe in Ontario whilst being cognisant of differences in institutional culture in countries such as Japan and the USA (Amati, 2008; Sorensen, 2002).

Arguably, the Green Belt has been very successful against its own objective of preventing urban 'sprawl' historically, as despite successive waves of deregulation and growing political recognition of the housing crisis, it has remained essentially untouched and impregnable as a policy, certainly until the Labour Government's proposals around 'grey belt' (Amati, 2007, p. 581; Dockerill and Sturzaker, 2019; MHCLG, 2024). The ideological debates between pro-development neoliberalism and anti-development localism within the Conservative Party, which has governed for most of the Green Belt's existence, are probably reflected in the wider country (Inch, 2015; Tait and Inch, 2016). Nevertheless, there has been an overwhelming popular and political desire to maintain England's 'green and pleasant' land, exemplified in the policy, despite Britain's liberal, capitalist economy. This enduring, unresolved pro- and anti-development contradiction is often worked out in the planning system, explaining why planning is often such a contested, conflictual space (Barry *et al*, 2018; Inch and Shepherd, 2019).

The chapter yields important insights informing other parts of the book. Firstly, historical institutionalism and how the past 'frames' the future, especially popular visions of history, shape and form the attitudes of planners, protestors and the public towards Green Belts (see Chapter 8) (Sorensen, 2015; Valler and Phelps, 2018, p. 1). Secondly, the Green Belt historically protecting the residential 'rural exclusivity' of the rural-urban fringe emerges strongly in literature and Chapter 6 will empirically examine the viewpoint that the Green Belt has been subject to 'regulatory capture' and 'rent-seeking' by homeowners (Cherry, 1996, p. 202; Sturzaker, 2010, p. 1014; Sturzaker and Shucksmith, 2011, p. 175). Moreover, whilst this chapter has explored the Green Belt's long-standing political nature (Monk and Whitehead, 1999; Sims and Bossetti, 2016), the *motivations* of campaigners and the public for supporting the policy is critically examined in Chapter 7, whilst Chapter 8 elucidates campaigners tactics. Chapter 9 draws these strands together by exploring ways that strategic planning could be rebuilt having charted its historic challenges and demise in this chapter. The Green Belt's history and longevity highlight how challenging reform of the policy is but also illuminate potential reforms (Hall *et al*, 1973A). The next Chapter (4) therefore explores the Green Belt's form/function and compares the Green Belt with urban containment policies internationally. This historical chapter is therefore anchored around the Green Belt's temporality, whilst the next one focuses on its spatiality.

Notes

1 By 1961, 27% of this purchased land was open space – much higher than the 3% of the current Metropolitan Green Belt which is publicly accessible (CPRE and Natural England, 2010, p. 2; Thomas, 1963, p. 18).

2 There was a gap between councils creating a *de facto* Green Belt through submitting an indicative Green Belt map to the Minister in the 1950s and the Minister approving the proposals due to post-war 'bureaucratic overload' (Cherry, 1996, p. 147; Sturzaker and Mell, 2016, p. 35). The Nottinghamshire Green Belt Map was first drawn in 1956 but

not approved by the government until 1989 following the 1980 Nottinghamshire Structure Plan (Hughes and Buffery, 2006, pp. 8–9). Although Manchester had an indicative Green Belt, it was not formally specified until the 1978 Greater Manchester Structure Plan and was approved in 1981. Green Belt approvals in other areas, like South Hampshire, did not take effect despite having their plan approved by the Minister in 1965 (Elson, 1986, p. 38).

3 Private sector estates built in the 1970s at Lower Earley, Berkshire, just outside the Metropolitan Green Belt, and South Woodham Ferrers, Essex, in the Metropolitan Green Belt, also inspired them (Ward, 2005B, p. 332).

4 A retired, long-standing West Midlands Conservative MP interviewed by the author gave this fascinating insight: 'Mrs Thatcher herself was a very staunch advocate of the Green Belt...I knew Nicholas Ridley well and, you see, a person's personal background can often influence them – he came from a landed family up in the north, but he was a very sensitive but...really quite a brilliant water colour painter'.

5 The Conservative Manifesto read 'we will protect and enhance the Green Belt' (Conservative Party, 2019, p. 31) and the Labour Manifesto was almost identical: 'we will protect the Green Belt' (Labour Party, 2019, p. 78).

6 This proposed urban extension in the Green Belt in places like Oxford and York (Manns, 2014; Mace, 2018).

7 Rees-Mogg has previously called for a national Green Belt review (Halligan, 2019).

8 For example, when the Planning Minister, Greg Clark, was drafting the NPPF (2011), he and the Chancellor of the Exchequer George Osborne were reported as contemplating relaxing the Green Belt (Manns, 2014, p. 16; Scott, 2015, p. 139). A fierce campaign by the *Daily Telegraph*, entitled 'Hands off our land', attacked the liberalisation of planning so the NPPF strongly upheld the status quo regarding the Green Belt with the NPPF's 5 Green Belt purposes copied verbatim from PPG2 (Mace, 2018, p. 3; DoE, 1995). The media reported that Savid Javid, Communities Secretary, proposed loosening Green Belt restrictions, especially on brownfield land, in the 2017 Housing White Paper and that Chancellor Philip Hammond proposed the same before the 2017 Budget. The opposition of the Prime Minister, Theresa May, saw off any proposed reform (The Economist, 2017).

4   The form and function of the Green Belt and an evaluation of its effectiveness

Introduction

Chapter 3 aimed to understand *how* the Green Belt in the UK became such a popular, well-established and geographically widespread policy compared to other countries, notwithstanding the economic and demographic growth pressures of post-war Britain (Rydin, 1985; Sturzaker and Mell, 2016). This chapter turns to *contemporary* discussions surrounding the policy by exploring its form *and* function with its fundamental *function* as an 'urban shaping device' explicitly aiming to shape urban *form* (Elson *et al*, 1994, p. 154; Gunn, 2007, p. 595). This contextualises the next section, which evaluates the numerous academic and think-tank critiques of the policy, alongside exploring the counterarguments and defences of it. In contrast to popular narratives about the Green Belt's success, this chapter evaluates literature which highlights its negative societal impacts, especially exacerbating the housing crisis (Hilber and Vermeulen, 2014). Finally, alternative policies and Green Belts internationally are evaluated to explore other examples of accommodating urban growth whilst protecting the countryside and environment (although these would be challenging to 'retrofit' in England (Oliveira, 2017)).

The Form and Function of the Green Belt

The Function of the Green Belt

The policy's overarching aim is 'to prevent urban sprawl by keeping land permanently open', and this containment objective is similar to 42/55 Circular and PPG2 (Department of the Environment (DoE), 1995, p. 5). Current national planning policy states its purposes as (Ministry of Housing, Communities and Local Government, 2024, p. 42):

a *'To check the unrestricted sprawl of large built-up areas;*
b *To prevent neighbouring towns merging into one another;*
c *To assist in safeguarding the countryside from encroachment;*
d *To preserve the setting and special character of historic towns; and*

DOI: 10.4324/9781032674315-4

e   *To assist in urban regeneration, by encouraging the recycling of derelict and other urban land'.*

It is, therefore, a regional *urban growth management* policy and *means* by which compact urban form can be achieved and sprawl prevented, rather than a blanket *countryside* policy or an end *in itself* as is popularly believed with surveys showing that 60% of people think that the Green Belt protects biodiversity and 46% that it protects areas of landscape quality (Nathan, 2007, p. 4). Moreover, while areas of biodiversity or environmental value *may be* in the Green Belt, it is not an *environmental* designation, such as AONBs (Areas of Outstanding Natural Beauty) or SSSIs (Sites of Scientific Special Interest), and its function is not determined by environmental characteristics (Barker, 2006; Healey, 2003; Mace *et al*, 2016). Likewise, although policy has attempted to improve the management of it to provide (more) open space and recreational opportunities (for example, DoE, 1995)[1], recreational/public accessibility is not its *primary purpose* (Kirby and Scott, 2023; Kirby *et al*, 2023A, 2023B, 2024; Mace, 2018). These clear aims and criteria can be seen in the proposed 'new' Green Belts/ Green Belt extensions that were not approved in the 1970s/1980s because they were deemed unnecessary (Table 4.1).

Nonetheless, the word 'green' in the name 'Green Belt' has captured the popular imagination, led the public to confuse greenfield with Green Belt and fuelled the belief that it is about protecting beautiful, scenic environments (House of

Table 4.1 Proposed Green Belts

Proposed **Green Belt**	Justification	Reason for refusal
Around Northallerton (North Yorkshire), between Worcester/ Malvern and Yeovil/ Sherborne.	Prevent coalescence between historic settlements/protect areas of landscape quality.	Unnecessary as the normal development management system was sufficient to restrict growth.
Tendring Peninsula (Essex) and around Scarborough/ Swindon.	Help control and restrict urban growth.	Unnecessary as low levels of housing demand and a concern not to 'devalue' the concept.
Extension of the Metropolitan Green Belt in Berkshire along to Reading.	Help control and restrict urban growth.	Refused so economic growth along the M4 'Corridor'/ housing growth in Central Berkshire could take place.
Oxfordshire/ Buckinghamshire County Councils sought to extend the Metropolitan Green Belt over the Chilterns to join the Oxford Green Belt in 1960.	Help control and restrict urban growth.	The government said it would 'devalue' the Green Belt concept.

Information from Elson (1986, pp. 19–21, 56–62) and Short *et al* (1987, p. 32).

Lords, 2016; Mace, 2018). Likewise, the policy has the underpinning concept of 'openness', which is fundamentally subjective, making assessment of the Green Belt challenging (DCLG, 2012; Mace, 2018). The other key characteristic, 'permanence', means that Green Belt boundaries can only be changed in 'exceptional' or 'very special' circumstances through passing a 'high bar' in local plans and planning appeals, so it is a 'blunt tool' and a strict, 'marginally flexible' land use designation (Abbott, 2013, p. 20; DLUHC, 2023, p. 44; Lloyd and Peel, 2007, p. 643). The policy has retained clear aims over time, and its strict application is remarkably geographically consistent across England. This is arguably symptomatic of Britain's centralised state, with no other Western capitalist country having such a direct and firm *national* constraint policy with international Green Belts reflecting city-wide concerns, for example, Seoul and Bangkok, or wider regional, state concerns, like Oregon or Florida (Boyle and Mohamed, 2007; Chu *et al*, 2017).

However, in practice, national infrastructure projects, like HS1/HS2; uses essential to a city's functioning and typically associated with the urban 'fringe', such as timber yards, sewage works, reservoirs and gravel extraction; certain employment types, like Nissan's Factory (Sunderland), Peddimore (Birmingham) and Cambridge's Science Park and ancillary agricultural buildings not fundamentally affecting the policy's 'openness' have all been approved in the Green Belt (Gallent *et al*, 2006, p. 457; Gant *et al*, 2011, p. 270).

The Form of Green Belts

Nationally, 1,638,150 ha of land is in the Green Belt, which at 12–13% of England's land surface, is similar to 1997 (Rankl *et al*, 2023, p. 1). By comparison, only 9–11% of England is urbanised, falling to 5–6% when excluding gardens and parks (DCLG, 2017, p. 9; House of Lords, 2016, p. 29). The initial reason for their establishment and urban form varies geographically according to local and regional circumstances (Table 4.2; Elson, 1986; Sturzaker and Mell, 2016). Scottish Planning Policy states that:

> For most settlements, a green belt is not necessary…**the spatial form of the green belt should be appropriate to the location**. It may encircle a settlement or take the shape of a buffer, corridor, strip or wedge
>
> (Cited in Manns and Falk, 2016, p. 15)

Green Belts can be characterised in form by their primary function, although these often overlap (Table 4.2).

Safeguarding the countryside from encroachment is a general objective applying to all Green Belts in England (Lloyd and Peel, 2007, p. 639). Likewise, assisting in urban regeneration is a general aim but more pertinent in areas with significant amounts of brownfield land, like Tyne and Wear and North Staffordshire (Foresight Report, 2010, p. 68; Gunn, 2007, p. 612). Some Green Belts are unusual in form as they are shaped by the sea, such as Avon (Longley *et al*,

Table 4.2 Form and function of the Green Belt

Function of Green Belt	Form of Green Belt	Examples
To check the unrestricted sprawl of large, built-up areas	These large Green Belts tend to be circular to prevent large urban areas from growing outwards.	Metropolitan, Avon, West Midlands, Nottingham, North Staffordshire.
To prevent neighbouring towns merging into one another	Can be a variety of shapes, from a 'buffer' between two settlements to a relatively 'blanket' Green Belt over a large area with lots of nearby settlements.	'Buffer' Green Belts: Worcester – Droitwich, Cheltenham-Gloucester, Burton-upon-Trent – Swadlincote. 'Blanket' Green Belts: North West, West Riding (Yorkshire).
To preserve the setting and special character of historic towns	Smaller but circular shaped to prevent outward expansion.	York, Oxford, Cambridge, Bath.

From Amati and Yokohari (2007, p. 125), Elson (1986, p. 39, 59–62, 176) and Munton (1986, pp. 207–208).

1992, p. 447). A Green Belt's size depends upon the area it is containing as having a 'proportionately' sized Green Belt was a key determinant in its implementation (Elson, 1986, p. 33). Naturally, the largest Green Belts are 'blanket' ones encircling large cities, whilst 'buffer' Green Belts are much smaller[2] (Munton, 1986, p. 208).

Evaluating the Green Belt

Debate on the effectiveness of Green Belts has been relatively continuous since their inception, although the first major critique was Hall *et al's* (1973A, p. 433) work, *Containment*, which argued that its effects, alongside that of the broader planning system, were 'perverse' and 'regressive' as creating a 'civilised version of apartheid' and an 'elitist … system distorted' which left 'the great mass of people betrayed' as 'those with the most…gained the most' (Hall, 1974, p. 386, 2002, p. 333). The debate has become increasingly intense recently with practitioner, academic and think-tank reports, like Policy Exchange and the Adam Smith Institute, calling for significant Green Belt reform (Table 4.6) (Airey and Doughty, 2020, pp. 39–42; Sturzaker and Mell, 2016, p. 42). Nonetheless, there are shared characteristics between critiques of the Green Belt and international critiques of planning systems.

Critiques of the Green Belt – Free Market Economic Perspectives

Critiques along economic lines can be broadly divided into wider philosophical arguments and more direct, empirical studies (Keeble, 1971).

Philosophical Arguments

The most general arguments against the policy relate to broader arguments against planning and state intervention in the economy (Cheshire, 2024; Cheshire and Buyuklieva, 2019; Hilber and Vermeulen, 2014). It is argued that settlements have always expanded or contracted in a 'natural', 'organic' way as a 'living organism' in response to market forces, agglomeration economics and people's individual preferences (see Chapter 3) (Lloyd and Peel, 2007, p. 640; Neuman, 2005, p. 14). Some academics, most notably Cox (2002, p. 3) and O'Toole (2007B), argue that planners/planning systems should not interfere with this 'natural' process of decentralisation and that, where they do through growth management policies, like Green Belts or Urban Growth Boundaries (UGB), the results are harmful to society (The Economist, 2014, 2017)[3]. Cox (2002, p. 3) highlighted that even 'compact' European cities have 'sprawled' since WWII, with Amsterdam's urban area growing by 65% and Paris's suburbs gaining 3 million people (Zonneveld, 2007). Others, such as Bruegmann (2005) and Layzer (2012, p. 152), argued that 'sprawl' reflects the desires of individuals and families for (more) domestic space, especially in the USA. The essence of these arguments is, therefore, if people desire to be homeowners, there are no sufficiently strong moral arguments to justify preventing people's freedom through planning restrictions and Green Belts (Breheny, 1997)[4].

Many of these libertarian arguments come from a North American perspective where public attitudes towards countryside conservation, land scarcity and car-dependency in the USA with its 'frontier mentality' being very different to Britain (Bunce, 1994, p. 25). Hall (2002, p. 332) argued that the American planning system was 'fairer' than the British one in that, in the absence of Green Belts, mass market housing was more affordable to lower-income families and had more domestic space. Hall (1974, p. 405) also underlined that rising land costs in the 1960s in Britain led housebuilders to raise the density of housing development, thus reducing its quality and permanently shifted their focus to more profitable, executive housing, although he conceded that the British system was more sustainable than the American one at protecting the countryside and promoting the efficient use of land.

Empirical Economic Arguments

OVERVIEW

Many academics, think-tanks and practitioners, whilst not advocating the abolition of the planning system or Green Belt, critique it *as a policy* and argue for its *reform* (Sturzaker and Mell, 2016).

The economic supply argument is essentially that by restricting the land supply when and where housing demand is highest on the edge of cities, the Green Belt raises land and house prices (Dawkins and Nelson, 2002; Evans, 1996; Monk and Whitehead, 1999). As Mace (2018, p. 1) argued, it often 'represents everything that is wrong with the English planning system…it is big state and woefully/wilfully ignorant of basic economics' while Hilber and Vermeulen (2014, p. 359) argued that England's planning system has 'ignored market signals and failed to cope adequately with changing socio-economic conditions'. The result, it is argued, is higher house prices, more cramped living conditions *within* cities and a dispersal/deflection of housing growth *beyond* the Green Belt ('leapfrogging') (Gallent and Shaw, 2007, p. 619). The policy is highlighted as a key reason why British property prices have grown rapidly compared to other OECD countries, and these countries have experienced higher levels of economic growth (Hilber and Vermeulen, 2014, p. 359).

EMPIRICAL DATA

Although the Green Belt has been successful historically in it *direct* aim of containing cities (see Chapter 3; Longley *et al*, 1992), it is now widely accepted by experts that it is *a* cause of Britain's housing crisis, although the *extent* of that responsibility is complicated and fiercely disputed (Barker, 2004, 2006; Cheshire, 2009; Evans, 1996; Lyons, 2014, p. 14). For example, Hilber and Vermeulen (2014, pp. 366–378) highlighted that, on average, newly built homes in Britain are 40% smaller than in the Netherlands, whilst their empirical study found that planning restrictions, like the Green Belt, raise house prices by around 30% in England and that they would have only risen 90% rather than 190% between 1974 and 2008 without these restrictions. Likewise, Ball *et al* (2014, p. 3010) found, in Melbourne (Australia), that house prices rose 56% (until 2008) after its UGB was implemented in 2002. Finally, Hall (1974, p. 404) highlighted that the price of land rose 20 times between the (late) 1930s and 1960s in Britain, showing the impact of the Green Belt on land values as, between 1892 and 1931, there was no significant increase in house prices, despite household numbers increasing 61% and incomes 31% (Figure 4.1).

The land market is key to the housing crisis because 70% of housebuilding costs result from the land price, an increase from only 4% to 12% in 1960, whilst residential land prices rose 1230% and house prices 500% between 1955 and 2008 although the UK's population only rose 21% (Cheshire, 2009, p. 11, 2014, p. 1, 2024; Manns, 2014, p. 9). It is argued, therefore, that price signals demonstrate that the policy is causing an 'artificial scarcity' of land for residential development[5] (Ball *et al*, 2014, p. 3010; Wolf, 2015A, 2015B). This land shortage is compounded by other planning restrictions across Britain, where 31% of the land is protected by AONBs and other designations, rising to 40% in the South East where housing demand is greatest (Nathan, 2007, p. 4; Saunders, 2016, p. 51). However, economic theory suggests that, whereas housing supply would normally rise in response to price/market signals to meet demand, like in the 1930s, the Green Belt prevents the market from 'clearing' (Cheshire, 2014, 2024; Dawkins and Nelson, 2002; Gilmore, 2016, p. 8). Nonetheless, whilst it is widely acknowledged that the housing crisis and land market *are* serious problems, establishing *direct* causal links between the Green Belt and high house prices is problematic due to the complexity of the housing crisis and difficulties in disentangling the *degree* to which the Green Belt is responsible compared to other factors, as even Hilber and Vermeulen (2014) acknowledge.

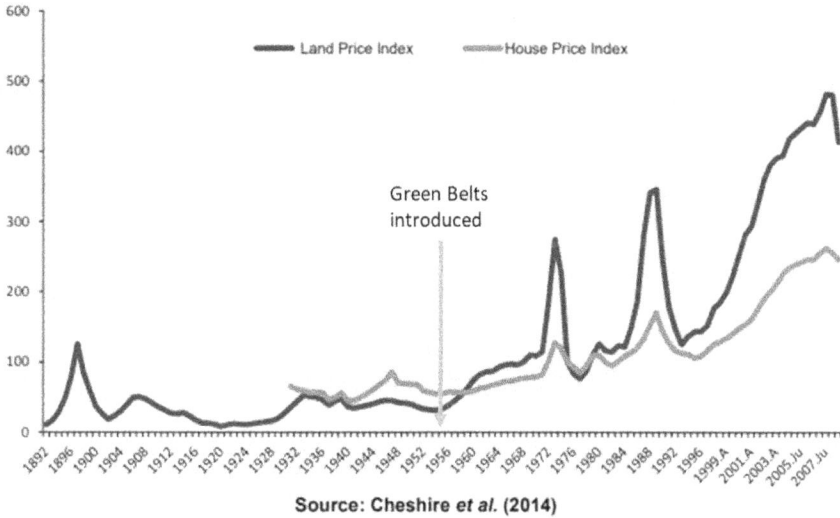

Source: Cheshire *et al.* (2014)

Figure 4.1 House price and land price increase between 1892 and 2007 (adapted from Coelho *et al* 2014, p. 23).

Defences of Green Belt – Critiquing the Free-Market Perspectives and the Land Supply Arguments

Firstly, Bramley (1993, pp. 1022–1024), based on empirical economic modelling of housing in 90 (English) districts, argued that house prices are largely determined by the 'second-hand' market. Consequently, even if a large programme of private housebuilding was initiated through widespread Green Belt release, the total housing stock would not increase by more than 1% annually and only have a 'very marginal/not very large' effect on prices[6]. More broadly, even the free market publication, *The Economist* (2024), argues that a large-scale new housebuilding programme by the Labour Government would not have a *significant* dampening effect on house prices. However, Bramley's model has been critiqued by Evans (1996), who contended that land release *does* reduce house prices, and Bramley (1993, p. 1037) conceded that Green Belt sites are 'disproportionately attractive' to housebuilders so may result in more housebuilding *ceteris paribus* compared to releasing other land. Secondly, Mulheirn (2019, p. 4) argued that housing supply has been keeping pace with population growth and therefore low interest rates, smaller household size, the 'financialisation' of housing and speculative mortgage lending have been driving high house prices recently (Ellis and Henderson, 2014; Saunders, 2016, p. 56, 150). Indeed, the 'porosity' of local housing markets to international/national demand and investment has been highlighted alongside the spatially variegated nature of the housing crisis (Bradley, 2023; Gallent, 2019; Meen, 2018, p. 489).

Bringing Perspectives Together – The Green Belt and Land Market

This goes to the heart of a complicated, multi-layered and contentious debate about whether house prices are *primarily* determined by supply or demand (the diagnosis) and if, after a significant land release in the Green Belt, housebuilders would build enough new homes to stabilise house prices (the prescription) (Ball *et al*, 2014; Evans, 1996; Hilber and Vermeulen, 2014; Saunders, 2016, p. 58)[7]. Meen and Whitehead (2020) argued that house prices are not perfectly elastic to supply because they are often determined by 'place' factors or locational characteristics, such as schools, amenities and environment, rather than *just* the quantity or quality of housing itself (this is known as 'imperfect substitutability' in Economics (Hilber and Vermeulen, 2014, p. 362)). Consequently, Pendlebury (2015) argued that releasing (more) land for housing in desirable places to live does not significantly reduce house prices because people still desire to live there, and it has been hypothesised that if the Green Belt were abolished, housebuilders would choose the most attractive sites to build executive housing rather than stabilising house prices (Monk and Whitehead, 1999; RTPI, 2015, 2016A; Saunders, 2016). Indeed, Whitehead *et al* (2015, p. 13), in a study of eight housing developments, and Bramley *et al* (2017), in research on housing developments in the South West (of England), both found limited impacts on local housing prices, especially if more and improved amenities resulted from development.

The counter-argument is that a significant release of land would dampen landowners expectations of 'hope' (land) value, send a clear signal to the market that 'artificial scarcity' is ended and stabilise house prices (Hilber and Vermeulen, 2014, p. 16; Mace, 2018). However, as the Economics Editor of the *Financial Times*, Martin Wolf (House of Lords, 2016, p. 35) argued, until a large-scale housebuilding programme takes place, it is difficult to accurately judge its effects on housing prices.

There *is* broad agreement about the oligopoly structure of the housebuilding industry but disagreement on whether releasing land from the Green Belt and more broadly would create a more 'contestable' market for SME housebuilders (Adams and Leishman, 2008, p. 9; Barlow and King, 1992, p. 380). The land market is characterised by huge competition between housebuilders, speculation by landowners and land promoters/traders and, consequently, an 'oligopoly' of large housebuilders whereby 50% of new houses built annually are by just the largest ten housebuilders (Archer and Cole, 2014; Bradley, 2021; House of Lords, 2016; Jefferys *et al*, 2015, p. 8).

Although beyond this book's scope, the fierce, wider debate about 'land banking', whereby housebuilders are accused of deliberately 'hoarding' or controlling land to 'drip feed' supply to keep prices high and maximise profitability, is clearly at the heart of this Green Belt debate (Kilroy, 2017, p. 9; Rydin, 2011, p. 120, 2013, pp. 120–121)[8]. This was investigated extensively by the Letwin Review (2018, p. 8), which found that limited housing output was caused by the local 'market absorption rate' whilst developers and consultants, such as Lichfields (2016, p. 20), highlight the long 'lead-in' times for sites to install utilities and infrastructure etc.[9]

Consequently, as argued in Chapter 6, more SME builders and integrated planning are required, with research showing that delivery rates are higher on sites prepared by public infrastructure investment, like Cranbrook (Devon) and Eastern Expansion Area (Milton Keynes) (Letwin, 2018; Lichfields, 2016, p. 6).

There is also the long-standing, contentious issue of land value capture and landownership, with the most radical Green Belt reform proposals calling for it to be nationalised and then sold to developers (McCrum, 2014), similar to the Netherlands, although less radical proposals call for a Green Belt 'tax' (Catney and Henneberry, 2019; Cheshire, 2009, p. 15, 2024; Wolf, 2015B, p. 2). This section demonstrates that private sector housebuilding *alone* on the Green Belt is not the 'magic bullet' solution to the multi-faceted housing crisis and that landownership and 'options' on land, taxation and the housebuilding industry's structure, including the wider difficulties which SME builders encounter (re)entering the housing market, all need to be considered (see Chapter 6) (House of Lords, 2016, p. 96; Lichfields, 2017, p. 16; Lyons, 2014, p. 21).

Defences of Green Belt – The Great Compaction Debate: Green Belt and Brownfield Land

There is a fierce, fundamental debate over whether sufficient brownfield land exists in British cities to meet housing needs (Mace, 2017; Shelter and Quod, 2016). For example, CPRE (2018, p. 10) regularly challenge the need to build on the Green Belt because they argue that there is enough brownfield land in cities to supply over 1 million new houses (excluding playfields/greenspace and including sites deliverable within 5 years). It is often argued and popularly assumed, that building on brownfield land is inherently more sustainable than greenfield as reducing the amount of land needed for development and benefiting from and supporting existing shops and services (Jenks *et al*, 2000).

However, as Mace *et al* (2016, p. 5) highlighted, these are often 'tricky', small sites unsuitable for large housebuilders, especially as many 'easier' brownfield sites have already been utilised, whilst several studies have questioned the amount of homes that brownfield land can accommodate, how long it can be sustained and the speed of delivery. The House of Lords Report (2016, p. 1398) calculated that less than 33% of the UK's housing needs until 2031 can be met through brownfield land. Abbott (2013, p. 13) and Young (2020) found that, although 100,000–150,000 acres of brownfield land are suitable for housebuilding, it is often in the 'wrong place' due to huge remediation costs (which limits the amount of affordable housing which can be provided). Moreover, housing delivery tends to be 50% slower on brownfield compared to greenfield sites (Lichfields, 2016, p. 18; Shelter and Quod, 2016).

Another problem with relying largely on brownfield sites is that, because they are generally in urban centres and inner-city areas, housing tends to be 1–2 bedroom apartments (Ball *et al*, 2014; Breheny, 1997, p. 207; Cheshire, 2014, p. 1, 2024). For example, 50% of dwellings constructed nationally were flats in 2007, whilst in Birmingham, 78% of dwellings built were apartments in 2018 (Best, 2019, p. 4; Cheshire, 2014, p. 1). While these apartments may cater well for young professionals and the elderly, most people (64% nationally), especially families with children, prefer having a house with a garden and living in quieter, semi-rural suburbs, with this trend reinforced by Coronavirus and the lockdowns (Breheny, 1997, p. 214; Goode, 2022B; Nathan, 2007, p. 5). Consequently, in a capitalist society where people (or at least those with resources) can choose where they live, people tend to 'vote with their feet' in the semi-rural suburbs (Mace, 2018).

Abbott (2013, p. 20) argued that the Green Belt has resulted in 'exoduses of (family) households' from the cities to more rural locations because of housing costs and the lack of housing choices (which was why some LPAs have allowed and justified executive housing developments in the Green Belt, such as Newcastle Great Park and Langley Urban Extension (Birmingham) (Birmingham City Council, 2017, p. 48; Gunn, 2007, p. 612)). The Green Belt has therefore been charged with causing 'town cramming' and is partly responsible for newly built homes in England having the least domestic space in Western Europe (Gunn, 2007, p. 598; Hall, 1974, p. 403).

Another key defence of the policy is that, as the NPPF (Ministry of Housing, Communities and Local Government, 2024, p. 42) stresses, it 'prevents urban sprawl' and 'assists in urban regeneration' and performs this function more effectively than alternative land use policies in other countries. A large body of literature, especially from North America, has critiqued sprawl for creating car dependency, congestion and more CO_2 emissions, increasing the inefficiencies of service provision and encouraging investors to abandon the inner-city (Davoudi and Sturzaker, 2017, p. 55; Gallent *et al*, 2006, p. 457; Layzer, 2012, p. 489). For example, Litman's (2015, pp. 4–40) survey of sprawl argued that it causes over $1 trillion in economic inefficiency, thereby highlighting a key benefit of England's Green Belt (see also Nathan (2007, p. 4)). Power and Haughton (2007, p. 164) called greenfield development 'land gobbling, congestion generating and environmentally damaging', so they call for a 'fixed UGB' (or Green Belt) to create compact, sustainable cities.

In turn, this feeds into the argument that the policy creates 'compact', walkable and sustainable cities which have been idealised by the concept of dense European medieval city cores (Breheny, 1997, p. 209; Jenks *et al*, 2000, p. 39; Neuman, 2005, p. 12). These arguments are used to support the Green Belt around historic cities like Oxford, York and Cambridge (Morrison, 2010, p. 166; Healey, 2003; Kells *et al*, 2007, p. 5). Although the North American emphasis is more on *managing* rather than *constraining* development, these arguments surrounding urban compaction are also similar to New Urbanist and 'smart growth' agenda arguments (Davoudi and Sturzaker, 2017, p. 59). Moreover, defenders of the policy argue that it makes housebuilders focus on 'harder' inner-city, brownfield sites rather than on 'easier' greenfield ones where they can maximise profit, especially in post-industrial cities with lots of brownfield land, like the Potteries, Tyne and Wear and Liverpool (Gallent *et al*, 2006; Monk and Whitehead, 1999).

In response, Neuman (2005, p. 15) has critiqued the underlying, 'fallacious' logic of the 'compact city', especially when society has changed so comprehensively since medieval times with modern transport. Secondly, as argued in Chapter 3, comparisons between urban growth and popular attitudes towards the countryside in Britain and the USA are largely unreasonable. Thirdly, many of those advocating Green Belt reform (i.e. Mace, 2018; Manns, 2014, p. 12), still argue for a 'brownfield' first policy.

Furthermore, many academics argue that the link between containment at the urban edge and inner-city brownfield development, enshrined and assumed in national policy, is highly contentious and unproven (Gallent and Shaw, 2007; Mace, 2017, 2018). For example, Gunn (2007, p. 612) and Monk and Whitehead (1999) argued that the resources devoted to urban regeneration are more important than the Green Belt in determining the amount of brownfield land developed. Indeed, arguably the tremendous popular fear of 'what if' regarding the Green Belt and campaigners/CPRE using the 'politics of affect' with the simple but powerful binary and 'dualism' of American-style sprawl *or* England's Green Belt, the countryside being 'concreted over' and already congested places becoming 'gridlocked' by development, shows the power of ideas regarding the policy (Sturzaker, 2010; Thrift, 2004, p. 57; Warren and Clifford, 2005, p. 361).

Critiques of the Green Belt – Environmental Perspectives

The policy is also critiqued because it is often not particularly 'green' or environmentally 'valuable' as it is largely used for intensive agriculture, which has net environmental disbenefits, whilst ecologically valuable land in it is usually protected by other environmental designations, like SSSIs, AONBs etc. (Gallent and Shaw, 2007; Prior and Raemaekers, 2007, p. 581)[10]. The Green Belt is frequently degraded near the rural-urban fringe as it is largely privately owned and often without public access or recreational opportunities, so it has a relatively low societal value (£889 ha) compared to urban parks and greenspaces (£54,000 ha) (Foresight Report, 2010, p. 98, 291). Indeed, the current 'brownfield first' policy has exerted development pressure on existing urban greenspace, with Cheshire (2014, p. 18) arguing that brownfield land often has a rich biodiversity, like the Hoo Peninsula[11].

Moreover, housing growth is often deflected, dispersed and redirected beyond the Green Belt, with many studies, like the Regional Studies Association's Report, *Beyond Green Belts* (Herington, 1990, p. 164), arguing that it has caused the unsustainable development pattern of 'leapfrogging' (Barker, 2006, p. 67; Town and Country Planning Association (TCPA), 2002). Firstly, this means that housebuilding sometimes takes place in the 'deeper' countryside, which often has higher ecological and environmental value than the Green Belt itself (Harrison, 1981, p. 114; Nathan, 2007, p. 5). Secondly, as areas beyond it often lack amenities and public transport links, they can become 'dormitories' and encourage car dependency, especially work, leisure and shopping trips *crossing* the Green Belt[12] (Morrison, 2010, p. 159). This not only increases pollution but also journey-to-work times, reducing time for leisure and family (Davoudi and Sturzaker, 2017, p. 62; Natural England, 2007, p. 2). Hall *et al* (1973B, pp. 49–55) argued that the policy was partly responsible for causing the 'suburbanisation' of the countryside and highlighted that CPRE's (then) position to the Barlow Commission was that 'controlled, peripheral development was better than leapfrogging and Green Belts'. Additionally, Nathan (2007, p. 5) and the Foresight Report (2010, p. 241) argued that it is 'better to allow managed expansion than unplanned leapfrogging'.

Counter Environmental Arguments in Defence of the Green Belt

Firstly, groups such as CPRE acknowledge that leapfrogging *has* occurred but usually argue that *planned* leapfrogging, especially new towns centred around public transport nodes and aiming to be self-contained, such as Bicester, Stevenage or Milton Keynes, are better than peripheral urban extensions poorly served by public transport[13] (CPRE, 2015).

Secondly, defences of the policy usually revolve around its recreational and environmental benefits (Kirby and Scott, 2023; Kirby *et al*, 2023A, 2023B, 2024), whilst Bradley (2019B) argued that it is well-used and patronised as local greenspace (Table 4.3).

As the policy is *not* a landscape designation, its recreational, environmental or landscape benefits are 'incidental' (Mace, 2018, p. 25). Harrison (1981, p. 109) found that, apart from a few well-known sites, it does not effectively serve the recreational needs of cities in the same way that local parks and urban greenspaces do, as seen in the lockdowns (for example, Mell, 2024; Mensah *et al*, 2016, p. 150; Roberts, 2017, p. 13).

The Green Belt is also argued to have general benefits related to green infrastructure more broadly, such as offering 'ecosystem services', providing 'lungs' and a carbon sink for cities, enhancing biodiversity/green infrastructure and reducing 'light pollution' (Amati and Taylor, 2010, p. 144, 152; CPRE, 2015, p. 4; Kells *et al*, 2007, p. 3; Kirby and Scott, 2023; Kirby *et al*, 2023A, 2023B, 2024). Again, these benefits are not incorporated into the Green Belt's purpose, whilst Mace (2018, p. 25) argues that it is disingenuous to link these wider benefits to inner-city and city centre areas with it mainly being *local* greenspaces which determine factors, such as air pollution.

Critiques of Green Belt – The 'Orphaning' of the Policy

Currently, there is not an alternative, 'counterbalancing' policy to provide for Britain's housing needs, with the Green Belt having been 'orphaned' as originally

Table 4.3 Recreational value of the Green Belt

30,000 km of public footpaths
Visited by 1.3 million people annually
Close to 30 million people live in cities
89,0000 ha of Green Belt land, which is protected by SSSI
Covers 33% of the UK's nature reserves
Covers 20% of England's woodlands

From Elson (1986), CPRE (2015, p. 2), Longley *et al* (1992, p. 449), Kells *et al* (2007, p. 3) and Thomas (1963, p. 23).

one of the post-war 'triptych' alongside regional policy and new towns, although the effectiveness of these arrangements was similarly critiqued (Hall, 1973A; Mace, 2018, p. 10; Papworth, 2015, p. 5). Indeed, when Circular 42/55 was issued, the architect of the 1947 Act, Lord Silkin, reflected:

> If the Government and local authorities are not able and willing to stand up to the pressures of agricultural interests, will not the result of the Green Belt policy be to induce local authorities to increase substantially the density of development...? The result [would] be that congestion and overcrowding will be as bad as...before and a new and greatly accentuated problem will be created for future generations...If that were to be the result of Green Belts, it might have been better not to have encouraged them after all.
>
> (Cited in Hall *et al*, 1973A, p. 57)

Likewise, in 1965, Hampshire County Council's Chief Planning Officer, Gerald Smart, argued that the policy is:

> Out of date. It is in danger of forcing the preservation of an archaic settlement pattern.
>
> (Cited in Hall *et al*, 1973A, p. 58)

Finally, Circular 42/55 was issued against the advice of civil servants, especially Sandy's Permanent Secretary, Dame Evelyn Sharp, who always maintained Green Belts were 'too big for comfort' and 'too rigid' (Cited in Sturzaker and Mell 2016, p. 35).

Defences of the Green Belt – Emotional and Institutional Arguments

As the Green Belt policy represents the *status quo* and enjoys widespread popular and political support, there has not been a pressing need to defend it until recently, resulting in limited academic or practitioner literature in favour of it (Amati, 2007; Parham and Boyfield, 2016). However, CPRE has consistently supported it and, with over 40,000 members, often employs deeply emotional and defensive arguments related to the English countryside, demonstrating the 'intensity of feeling' associated with the 'politics of affect' (Amati, 2008, p. 2; Lloyd and Peel, 2007, p. 640; Thrift, 2004, p. 57). Moreover, it is argued that the policy's benefits cannot be exactly quantified by monetary value as being more intangible, whilst some things are inherently worth protecting because of their intrinsic value and should not be spoiled by profit-making, capitalist activity, such as the countryside or heritage (Amati, 2008, p. 4; Matless, 1998, p. 32; Rydin, 2011, p. 7 – see also the political ecology literature on the capitalisation and exploitation of nature and greenspace (for example, Malik, 2024; Walker, 2007)). Indeed, as outlined in Chapter 3, the policy's historical institutionalism and iconic status means that it appears self-evidently a 'right' or 'natural' policy (Mace, 2018; Munton, 1986) (see Table 4.4).

Table 4.4 Quotes about the importance of the Green Belt

An 'iconic and popular planning policy' capturing the countryside's 'popular image' (Amati, 2007, p. 580).

The Green Belt is a 'sacrosanct', 'immovable shibboleth' which has 'iconic status', a 'baggage' of values and longevity (Prior and Raemaekers, 2007, p. 596).

'Widely supported' as the 'first article of the planning creed' (Gant *et al*, 2011, p. 267).

'Hallowed by popular support and fears of what would happen if they are weakened' (Cullingworth *et al*, 2015, p. 183).

The 'cornerstone' of planning (Amati and Yokohari, 2007, p. 311; Amati and Taylor, 2010, p. 152).

The 'very *raison d'être* of planning' – Desmond Heap, TPI President (1955) (Cited in Amati and Yokohari (2007, p. 315) and Mace (2018, p. 25)).

The 'magic weapon'/'mighty sword in planning's armoury (Kells *et al*, 2007, p. 1; Law, 2000, p. 37).

'One of the few planning concepts which has some meaning for the man in the street... much loved...as a means whereby 'they' (the authorities) will protect the open country around the city'. A 'statutory shibboleth respected by planners, politicians and the public' (Law, 2000, p. 57; Lloyd and Peel, 2007, p. 645).

'One of the most internationally famous attempts to control urban growth', part of a 'universal planning cannon' and a 'central plank of national planning policy' (Amati, 2008, p. 1, 2).

The 'iconic pillar of post-war planning' (Parham and Boyfield, 2016, p. 10).

Chapters 7 and 8 chart the extent of popular support for and misunderstanding of the policy, but the statistics in Table 4.5 demonstrate why it is politically 'toxic' and 'untouchable' to reform as often likened to a political 'sacred cow', like the NHS (Munton, 1986, p. 211).

Table 4.5 Statistics on Green Belt's popularity

Key area	Statistics	
Popularity	66% of people surveyed by CPRE (2015) want the Green Belt to be protected.	Foresight Report (2010, p. 60) recorded 85% of people supporting it.
Misunderstanding	71% of CPRE's (2015) respondents said that they knew 'little to nothing' about its purpose.	Foresight Report (2010, p. 60) found that most people think GB protects wildlife (only 25% that it contained cities).
Temporality	Support for the GB has fallen in CPRE's survey from 80% (2005) to 60% (2015).	
General Views	66% of people think that 25% of England is urbanised when only 9–11% is.	

Information from Amati (2008, p. 20), CPRE (2015, p. 1), Foresight Report (2010, p. 60), House of Lords (2016), Mace (2018), Papworth (2015, p. 65) and Whall (2015, p. 29).

Alternatives to Green Belt Policy

Planning without the Green Belt

Hall *et al* (1973A, pp. 379–385) hypothesised that, without the post-war Green Belt, London's peripheral growth would have only continued 2–3 miles further than its current boundary as confined by the Underground's reach (12–13 miles). However, freestanding towns with good transport links in the Green Belt, like Redhill and St Albans, would have expanded (see also Sturzaker (2011, p. 556) and Warren and Clifford (2005, p. 356)). Moreover, councils, especially in London and Birmingham, would have continued to purchase land for the Green Belt, like Robert Moses's parks in New York, and North American-style sprawl would not exist (Hall, 1974, p. 417, 2014, p. 198).

The Garden City Alternative

With managed but continuous 'cellular' growth, Hall *et al* (1973A, pp. 387–389) predicted that the close proximity but separate nature of Welwyn, Letchworth and Stevenage to each other is what the Home Counties would have looked like. Hall likened this spatial configuration to Los Angeles.

Other Alternatives: Historic and Modern

DEVELOPMENT IN A PLANNED WAY: TRANSIT ORIENTED DEVELOPMENT (TOD)

Most proposals for reforming the Metropolitan Green Belt revolve around TOD (Table 4.6) (Haywood, 2005; Manns, 2014). TOD is essentially developing housing and services around transport hubs as a more sustainable alternative to car-dependent development (Carlton, 2007; Goode, 2023). It is broadly tied to the New Urbanist agenda of more traditional, high-quality design and walkable neighbourhoods facilitating social interactions (Menotti, 2005, p. 111). The most famous historic TOD-led plan is the 1947 'Finger Plan' of Copenhagen (Papa and Bertolini, 2015, p. 70).

Nevertheless, TOD around railway stations in Britain has been limited (Haywood, 2005, p. 89; Mace *et al*, 2016, pp. 6–32). For example, Warwick Parkway Station and stations, like Iver and Taplow on the Elizabeth (Crossrail) Line, have frequent transport connections yet do not have housing developed around them due to the Green Belt (Bristol Parkway is one exception) (Cheshire, 2014, p. 2, 2024; Haywood, 2005, p. 88). Indeed, an RTPI (2016B, p. 6; 2024) study, *The Location of Development*, found that just 13% of new housing developments were within 800 m of train stations between 2012–15, 20% between 2015–17 and 17% on average. Most proposals for Green Belt TOD therefore propose 'ped-shed' developments within 800 m of train stations and exclude land of high environmental, agricultural or recreational quality (Cheshire, 2014; Goode, 2023, p. 2; Mace, 2018, p. 15). The main underground network, which extends into the Green Belt, is also being studied, with places such as Theydon Bois and Chesham suggested as sustainable

Table 4.6 Proposals for development on the Metropolitan Green Belt (MGB)

Study	Proposal
Cheshire (2014, p. 2)	Build on less valuable Green Belt (GB) land within an 800 m radius (or 10-minute walk) of railway stations totalling 1 million homes and taking up 1% of total GB land.
Papworth (2015, p. 3)	3 million homes within a 2.5 km of stations, taking up 0.5% of total GB land.
London First/Quod (2015, p. 20)	3 million homes within a 2.5 km of stations, taking up 3% of total Green Belt land.
Clarke *et al* (2014, p. 3)	2.5 million homes within a 2 km radius of stations, taking up 5% of GB land.
Royal Town Planning Institute (RTPI) (2015, p. 2)	Cheshire's (2014, p. 2) proposals would result in 4–7.5 m more car journeys as only 7.4% of commuters in the sample of 5 MGB towns, Bracknell, Maidenhead, High Wycombe, Watford and Hemel Hempstead, commuted by train to London.
Stringer (2015, p. 1)	20,000 ha of non-environmentally protected land in the MGB within 800 m of stations.
Cheshire and Buyuklieva (2019, p. 1)	Build within 800 m of stations with a 45-minute journey into London. Introduce a Land Development Charge of 20% and Green Development Corporations through granting development rights to the railway industry.

See also Holman *et al* (2015, p. 18) and Mace *et al* (2016, p. 39).

development locations (Mace, 2018). The aim is to uphold the Green Belt's *general* purpose and protect valuable land while meeting London's pressing housing needs in a (more) sustainable way. As London has a high-quality public transport network, it has been the most studied in terms of TOD proposals (Kilroy, 2017; Table 4.6). Nevertheless, Peter Brett Associates Study (2015, p. 40) found that Birmingham's housing shortfall of 38,000 could be accommodated in a 1200 m radius of just *four* stations in the West Midlands Green Belt (Whitlock's End, Blakedown, Blake Street and Shenstone) (see also Goode (2023)).

However, TOD generally, alongside TOD in the Green Belt, has been heavily critiqued (O'Toole, 2007A). Firstly, even in the Metropolitan Green Belt, most trips are not commuting into central London but either locally or to other towns/ cities, mainly by car (RTPI, 2015, p. 7). Moreover, other trips are taken alongside the work commute for increasingly complicated patterns of recreation and shopping (Budnitz *et al*, 2020). Consequently, an RTPI Report (2015, p. 12) found that Cheshire's TOD plans would result in 4–7.5 million more car trips (see Table 4.6). This trend would be magnified in regional Green Belts, where much less capacity exists on public transport (Goode, 2023). Moreover, Mace (2018, p. 17) argues that TOD would be politically challenging because planning battles and protests would be likely over each TOD proposal in the Green Belt, whilst this selective release of land would not send a clear enough signal to land markets to significantly reduce prices. Finally, TOD would potentially undermine the Green Belt's strategic integrity, leading to a 'beads on a string' form of development (Mace, 2018, p. 15).

Green Wedges, Webs and Corridors

The main land use policy alternatives of green wedges, webs and corridors, which have been implemented around the world, are now explored (Oliveira, 2017)[14].

Green Wedges

History

Green wedges or 'ducts of greenspace from the countryside into the centre of a city' have long been advocated by planners in many countries and are sometimes proposed as alternatives to Green Belts (Oliveira, 2014, p. 357, 2017). Mace (2018, p. 5) highlighted how, in 1829, John Claudius Loudon proposed a London Plan including green wedges, whilst Oliveira (2014, p. 362) explored how green wedges were supported at the 1910 RIBA Town Planning Conference, most prominently by the architect Vaughan Lanchester, who contended that:

> The prevalent view is that parks and recreation grounds should form a ring round the city; it is, however, **difficult to see on what basis this view rests… the parks themselves should be place radially.**
> (Cited in Oliveira, 2014, p. 360)

At the same Conference, the famous engineer Rudolf Eberstadt argued about Green Belts that:

> *It is injurious and hurtful to town expansion…we must break down the ring. The pattern for modern town expansion is the radial pattern.*
> (Cited in Oliveira, 2014, p. 362)

Abercrombie (1944, p. 103) was influenced by these ideas and sought to combine green wedges *and* Green Belts in the *Greater London Plan*. However, due to complications in land use and ownership with green wedges, they were not implemented apart from the Lee Valley, unlike Birmingham, with its long history of green wedges (Oliveira, 2014, 2017; Sturzaker and Mell, 2016). Indeed, whereas the concept of green wedges largely receded in the popular imagination, the Green Belt has become incredibly popular, demonstrating the 'leitmotif of the power of planning ideas' (Morrison, 2010, p. 166; Oliveira, 2014, p. 369).

Benefits and Disadvantages

Herington (1990) recommended green wedges over Green Belts as a better way to manage urban growth. Firstly, they provide greenspace directly to more people in urban areas and successfully connect city centres to the countryside literally and psychologically (Amati, 2008, p. 11; Lloyd and Peel, 2007, p. 642). Moreover, wedges have become more popular recently with the rise of ideas, like 'green' infrastructure/grids, which underline the benefits of *integrating* rural and urban (Gallent and Shaw, 2007, p. 617; Prior and Raemaekers, 2007, p. 594).

Wedges can follow linear routes, such as canals and rivers, so they are better at facilitating recreation (Amati, 2008, p. 13). Secondly, they are flexible and 'more sophisticated' as they are able to expand with urban growth and be adjusted and altered in response to changing economic and social forces more easily than 'rigid' Green Belts (House of Lords, 2016, p. 86; Scott *et al*, 2013, p. 13, 40). For example, the green wedges of Copenhagen's 1947 'Finger Plan' have grown with its outward expansion, and, although in response to increased housing demands, the 'fingers' of development have become 'fatter', they have not joined up (Davoudi and Sturzaker, 2017, p. 69; Oliveira, 2014, p. 369). The (then) Chair of Natural England, Sir Martin Doughty, argued in 2007 that:

> The time has come for a greener green belt. We need a 21ˢᵗ century solution to England's housing needs which puts in place a network of green wedges, gaps and corridors, linking the natural environment and people.
>
> (Cited in Scott *et al*, 2013 p. 14)

However, the Green Belt's primary purpose is not recreation but to *prevent urban sprawl,* which green wedges fail to do (CPRE, 2005). Moreover, due to existing settlement and land ownership issues, green wedges are often difficult to implement retrospectively (Oliveira, 2017).

Green Web

The Green Belt's principal aim, in common with modernist planning ideas, was to separate town and country, whereas the green web concept is premised on the benefits of *integrating* them (Manns and Falk, 2016, p. 18). As the 'web' analogy suggests, it proposes a more complicated, imaginative and flexible mix of housing, transport and greenspace than the Green Belt's traditional, neat separation (Yokohari *et al*, 2000). Given West London's severe housing and development pressures, Manns and Falk (2016, p. 6) recommended a 'green web' there[15]. Moreover, Amati (2008, p. 9) explained the failure of Green Belts in Southeast Asia as directly 'copying' Britain's Green Belt, whereas Asian vernacular development has traditionally integrated agriculture and paddy fields into cities.

Green Corridor

This concept weaves the green wedge and web ideas together, although it has been largely developed with the existing Green Belt in mind (Mace, 2018). Essentially, it involves designating a large corridor where housing and industrial development is allowed to take place, often around TOD, but where access to and facilities in undeveloped Green Belt is improved through planning gain (Amati and Yokohari, 2007; Mace, 2018). The corridors are planned strategically and aim, through signalling to the market that a large amount of land is going to be released to dampen land prices (Scott *et al*, 2013). This has been recently advocated by Mace (2018, p. 18) based on the *London Plan's* coordination corridors. However, Mace *et al* (2016, p. 7) argued that, initially, a 'pioneer corridor' should be trialled on the existing London-Cambridge Corridor to build public confidence about Green Belt reform. Finally, Mace's ideas have been extended by Manns (2014, p. 14), who proposed much larger and longer corridors in the Greater South East such[16].

Other Green Belts Around the World

There are examples internationally of Green Belts or UGBs, but where they exist, generally, the aim is to *shape and manage* urban growth, such as in Portland and Seoul, rather than *prevent* it with more autonomy for regions and cities (see Table 4.7 and Chapter 10) (Coelho *et al*, 2014, pp. 35–39; Daniels, 2010, pp. 259–261).

Key Findings from the Form, Function, Evaluation and Alternatives Chapter

This chapter's aim has been to examine the Green Belt's current form and function whilst evaluating its effectiveness and exploring alternative policies. It has shown how inextricably linked the Green Belt's form and function are, thus underscoring the importance of space in researching the policy. That the Green Belt still commands such widespread popular political support and emotional appeal when England is experiencing a serious housing crisis and it is widely attacked

Table 4.7 Green Belts around the world

City	Urban policy description
Tokyo	Green Belt: introduced in 1956; abolished in 1969 due to pressure from landowners.
Christchurch	Green Belt: Introduced in the 1950s; abolished by the Resource Management Act (1991)
Seoul	Green Belt: Introduced in 1970 Urban Planning Act – flexibly applied as land is frequently released for housing.
Bangkok	Green Belt: approved in 1971 – not continuous and away from the urban edge to allow urban expansion. Consists of 700 km^2 of paddy fields to prevent flooding.
Sydney	Green Belt: Introduced in 1945; abolished in 1960 due to landowner opposition.
Melbourne	Urban Growth Boundary: Introduced in 2003; relaxed in 2008.
Portland	Urban Growth Boundary (UGB): Introduced in 1979 – partially repealed by Measure 37 (2004) and then partially reinstated by Measure 37 (2007). UGB is still in place.
Durban	Informal Green Belt introduced.
Medellin	Green Belt: Introduced in 2012 to protect its hill slopes from mudslides.

Information from Amati (2008, pp. 11–13), Amati and Yokohari (2007, p. 315), Amati and Taylor (2010, p. 152), Ball *et al* (2014), Boyle and Mohamed (2007, pp. 681–692), Coelho *et al* (2014, pp. 35–39), Chu *et al* (2017, pp. 383–384), Daniels (2010, pp. 259–261), Gant *et al* (2011) and Layzer (2012, pp. 500–505).

as a policy is remarkable and builds on Chapter 3, which charted how the Green Belt acquired such a prescient place in the popular imagination (Mace *et al*, 2016). Additionally, the literature suggests that certain groups, especially homeowners, are particularly powerful in the planning system, especially in support of the Green Belt (Sturzaker and Mell, 2016), so the next theoretical Chapter 5 critically explores power and interest groups in planning.

Notes

1 The 2023 NPPF explained that 'local authorities should plan positively to enhance their beneficial use, such as looking for opportunities to provide access; to provide opportunities for outdoor sport and recreation; to retain and enhance landscapes, visual amenity and biodiversity; or to improve damaged and derelict land' (DLUHC, 2023, p. 44). The 2019 Conservative Party Manifesto (p. 31) included the aim of enhancing the Green Belt.

2 The largest Green Belt is the Metropolitan Green Belt at 516,000 ha – three times the size of London's urban area and 3.8% of England's land surface (Mace *et al*, 2016, p. 4; Munton *et al*, 1988, p. 3). Other blanket ones are about half the size of the Metropolitan Green Belt but a similar size to each other with the West Riding (248,241 ha), North West (247,650 ha) and West Midlands (225,000 ha) (CPRE and Natural England, 2010, pp. 20–25; Munton, 1986, p. 212). 'Buffer' Green Belts are smaller, with the Burton upon Trent/Swadlincote one 714 ha and Cheltenham/Gloucester one 6,694 ha (CPRE and Natural England, 2010, p. 14).

3 O'Toole (2007B, p. 1, 93) attacked planning as 'coercive' arguing that it 'harms your quality of life and... future', while Cox (2002, p. 3) defended urban 'sprawl' as the 'world's oldest land use trend' and argued 'I favour freedom, and no compelling justification has been demonstrated which justifies the abridgement of freedom necessary to outlaw urban sprawl'.

4 There are equally strong moral arguments in favour of planning systems and the Green Belt, especially related to the societal disbenefits of urban sprawl (Litman, 2015; Davoudi and Sturzaker, 2017, p. 55). For example, Crook (2015; House of Lords, 2016, p. 1232) argued that a managed release of Green Belt land is better than an unplanned release as then planners can use their place-making skills and planning gain to coordinate the facilities/infrastructure needed alongside housebuilding (Holman *et al*, 2015, p. 9).

5 For example, agricultural land in Oxfordshire has a value of £10-25,000 per ha, whilst land with planning permission for residential development is worth around £5.6 million per ha (Bradley, 2021, p. 395).

6 Bramley's (1993, p. 1022) found that even a large release of land from the Green Belt would only increase housing output by 2% and reduce prices by 1.2%. He argued that state intervention in the land market to purchase land for housebuilding and providing social housing would be a more effective way to solve the crisis.

7 These questions are directly explored in the empirical Chapter 6, but an overview is given here.

8 A 30–40% gap nationally between annual permissions and build-out rates is pointed to. There were 850,000-1 million unbuilt permissions in 2018 (Mace, 2018, p. 12; Bradley, 2021, p. 394).

9 Letwin's findings echo consultants reports on urban extensions, with Peter Brett Associates (2015, p. 14) finding that housebuilders are limited to building around 106 units annually and Lichfields (2016, p. 6) 161 units. 'Land banking' is argued to not be in the business model of housebuilders with shortages of construction workers/materials and traditional, time-consuming methods underlined (Gilmore, 2014, 2016; Lichfields, 2018)

10 Sixty percent of the total Green Belt, 74% of it around Cambridge, 71% around Birmingham and 60% around London, is intensively farmed (Amati and Taylor, 2010, p. 144; Cheshire, 2014).

11 Between 1992 and 2005, 50% of London's and 600 acres of England's playing fields were used for housing, and the equivalent of 22 Hyde Parks were developed from front gardens for driveways (McCrum, 2014, p. 2).

12 The A40 carries up to 32,000 vehicles per day between Oxford and Witney, whilst the A14, between Huntingdon and Cambridge, carries 85,000 (Highways England, 2018, p. 1; Self, 1962, p. xv).

13 A planner at an environmental charity interviewed argued that: '*New towns were generally well planned...put next to railway lines...* [as] *there were homes and jobs in those areas, they were reasonably well self-contained*'.

14 Green gaps are not explored here because they are a more local rather than a city-wide or regional growth management policy (Elson, 2002; Natural England, 2007, p. 2; Scott *et al*, 2013, p. 13).

15 In the All Party Parliamentary Group for London's Planning Report, the Chair Rupa Huq MP endorsed the green web plan calling the Green Belt 'King Canute' as 'resisting the tide' (Manns and Falk, 2016, pp. 36-46). The plan involved 200,000 new jobs and homes in the 'Western Wedge', a 'garden city' in the Green Belt at Northolt, a West London orbital railway, a 'green web' centred on the Colne Valley and a 'Blue Corridor' on the Grand Union Canal (Manns and Falk, 2016, p. 2–6).

16 From Hammersmith through Reading to Newbury (west), Romford through Chelmsford, then Colchester (east), Greenwich to Ashford (southeast) and Barnet to Milton Keynes (northwest) (Manns, 2014, p. 14).

5   Power and interest groups in planning

A theoretical frame

Introduction

This chapter develops the book's broader theoretical framework through exploring wider planning and geographical theories in the international literature and critically evaluating the extent to which they can be applied to the Green Belt (Allmendinger, 2009; Prior and Raemaekers, 2007). Firstly, it explores the Green Belt's popularity and permanence through 'theoretically informed critique' and analysis of *which* groups have the most power in planning and *how* this power is exercised (Parker *et al*, 2015, p. 520). Secondly, it uses the policy as a lens or medium to develop a deeper understanding of power and interest groups in planning and wider society. Rydin (1985, p. 7) argued that this is the ultimate aim of planning research. Thirdly, it contributes to and refines aspects of planning theory by critically assessing to what extent they apply to the Green Belt.

The chapter examines what Healey *et al* (1988, p. 152) highlighted as the three main conceptual frameworks for viewing interest groups in planning systems: positivist, Marxist and post-structuralist. Firstly, it finds that the positivist approach is useful for identifying interest groups but fails to account for the uneven power relations in planning (Flyvbjerg, 1998; Fox-Rogers and Murphy, 2014). Secondly, while aspects of Marxian theory are useful, through focusing largely on city centres and the inner-city, it fails to take sufficient account of and explain power relations at the (under-researched) 'rural-urban fringe' – this book's spatial focus (Gallent and Shaw, 2007, p. 620; Scott *et al*, 2013, p. 1). The Green Belt is a prescient example of the state *restricting* capital accumulation by volume housebuilders and, rather than a conflict between labour and capital, can be conceptualised in a more sophisticated way as the arena where the fierce 'intra-capitalist' conflict between homeowners and housebuilders is played out (Foglesong, 1986, p. 22; Lake, 1993, p. 89). The book utilises this concept of an 'intra-capitalist' conflict as the main theoretical framework (Kiernan, 1983, p. 72). Finally, a post-structuralist view of planning and the academy's response in theories, like collaborative planning, also fails to account for certain groups, such as homeowners and housebuilders, having *more* power than others (Tewdwr-Jones and Allmendinger, 1998).

This conceptual framework is operationalised through the social justice or just city approach as an analytical framework to evaluate the Green Belt (Fainstein,

DOI: 10.4324/9781032674315-5

2010, 2014). This approach, essentially evaluating 'who benefits and loses' from planning (Kiernan, 1983, p. 83; Sandercock and Dovey, 2002, p. 152), goes back to Hall *et al's* (1973A, 1973B) work but has been associated more recently with Fainstein (2010), Soja (2010) and Flyvbjerg's (2004) work. The chapter highlights that any proposal for Green Belt reform needs to be evaluated by *political and social acceptability* alongside economic and practical viability (see Chapters 3/4; Breheny, 1997, p. 209). The chapter then explores the political nature of planning systems by examining *how* powerful groups, especially homeowners, attempt to exercise power (Altshuler, 1966; Cherry, 1982). It employs Luke's (2004, p. 14–59) *Three Dimensions of Power* to elucidate how the workings of power are not always overt but are often more subtle. It looks at *processes* and *outcomes* which traditionally have been explored separately in planning research (Fainstein, 2014; Rydin, 1985).

The chapter aims not only to gain a deeper understanding of planning but, in line with calls in Geography for 'policy relevant' research, to help inform solutions to the housing crisis and ensure that planning outcomes are more just (Dorling, 2015; Martin, 2001; Ward, 2005A, p. 310). Nevertheless, understanding power is still vital to shape policy (Flyvbjerg, 1998), so the chapter feeds into the methodological approach by shaping the analytical framework and research methods used.

Theoretical Frames of Interest Groups and Power in Planning: Towards a Conceptual Framework of the Green Belt

Healey *et al* (1988, p. 152) helpfully identified three main conceptual frameworks for viewing interest groups in planning: positivist, Marxist and post-structuralist. However, arguably none of these theoretical approaches are sufficient at explaining the system's complexity or accounting for the Green Belt's popularity and longevity.

Positivist or Rationalist

This viewpoint essentially casts planners as apolitical, 'technical, disinterested and even-handed' and assumes that people have access to perfect planning information and make 'rational' decisions (Kiernan, 1983, p. 72). Indeed, local people are viewed as dispassionately and objectively weighing up the costs and benefits of particular developments whilst, in public consultation, the views and concerns of communities are well-represented and, if people have not expressed a view, this represents their 'tacit support' (Healey *et al*, 1988, p. 154; Rydin, 1985, pp. 9–11). Likewise, local politicians have a good knowledge of their communities and generally represent their 'best' interests (Yarwood, 2002, p. 277). Consequently, planners and politicians are said to make decisions in the public interest as weighing up and understanding public opinion (Sturzaker and Shucksmith, 2011, p. 189).

This approach has been criticised as neglecting the key issue of power in planning (Flyvbjerg, 1998; McLoughlin, 1985). As early as the 1960s, Altshuler's (1966) work in the USA highlighted that planners were restricted in what they could achieve by political constraints and realities. He characterised planning as more about *managing* competing and conflicting interests. More recently, Rydin and Pennington (2000, p. 154, 156) argued that planning is liable to 'special interest capture' because of 'selective participation by vocal and well-organised interest groups', such as homeowners, leading to policies which benefit particular groups while the '*costs of policy failure spread across non-mobilised sections of the community*'.

Indeed, a large body of research internationally has explored the political nature of planning systems and found that consultation tends to be dominated by the 'usual suspects' of 'seasoned campaigners' – older, middle-class homeowners with plenty of time who are 'ferocious and articulate' but not necessarily representative of their communities (Sturzaker, 2011, p. 566; Table 5.1). Simmie (1981), in his study of Oxford, found that, in planning, 'unorganised groups...come off badly, bearing regressive cost' (cited in Cherry (1982, p. 115)). This underlines the importance of social capital in planning systems and community engagement. For example, Wills (2016, pp. 43–62), in case studies of Neighbourhood Plans in St James (Exeter), Highgate (London) and Holbeck (Leeds), highlighting that those involved tended to be 'older, wealthier, better educated, long-term residents' living in areas with higher levels of 'civil infrastructure, core and capacity'. The positivist model arguably underestimates the challenges of involving the majority of people

Table 5.1 Studies on 'objector' groups in planning

Study	Location	Findings
Hubbard (2005, p. 3, 12)	Asylum centres in Oxfordshire/ Nottinghamshire	Often refused because of local campaigns and the countryside being 'a repository of white (middle-class) values, ideologies and lifestyles'.
Davison *et al* (2016, 2017)	Sydney, Australia	Politicians can exploit community opposition to affordable housing as part of campaigning strategies and for political advantage.
Sturzaker (2010, p. 1014, 2011, p. 566)	Long Compton (Warwickshire), Rainton (North Yorkshire) and Baselow/Thorpe, Derbyshire	Lack of rural affordable housing due to snobbery and the 'exclusionary preferences of the powerful'. A local council representative labelled objectors 'retired, professional, middle class' while a housing provider called them 'a very articulate group of people'.
Kjærås (2024)	Oslo, Norway	Explore the politics of urban densification.
Sims and Bossetti (2016, p. 20, 27)	Outer Boroughs of London	One Borough Director of Planning is cited as saying that there is a 'handful of articulate, well-resourced residents who will oppose anything'.
Short *et al* (1987, p. 37)	Central Berkshire	88% of people forming anti-development, 'stopper campaign groups', were homeowners.
Ball's (2004, pp. 121, 132)	Urban regeneration projects in London	Survey of housebuilders and councils found that 47% said the consultation process was 'unrepresentative', 'dominated by a small group' and 'undemocratic'.

See also Warren and Clifford's (2005, p. 370, 378) study on St Andrews.

in planning systems, including time, intimidation by planning's official, technical nature or wider disengagement with local democracy (Lowndes and Pratchett,
2012, p. 453).

The positivist concept of perfect information and dispassionate rationality is
also problematic because new housing development regularly represents an immediate, clear and direct locational 'threat' to local residents, which often unites them
strongly together (Gant *et al*, 2011, p. 268; Sims and Bossetti, 2016, pp. 20–27).
This 'threat' includes the prospect of lower house prices, harm to local amenities,
fear of change/outsiders and the 'negative externalities' of development, such as
extra congestion and strain on services (Nathan, 2007, p. 4). Rather than being an
objective system as premised by the positivist planning model, Coelho *et al* (2014,
pp. 3–7) argue there are 'asymmetries of power' and 'disproportionate opportunities for small groups to block development' meaning that the system is 'distorted
in favour of homeowners'. However, the benefits of more affordable housing are
long(er) term, less tangible or direct, and those likely to benefit, such as renters and
adults living at home, may not be aware of development or its benefits due to the
lack of perfect information, and form a more disparate, geographically dispersed
group (Whitehead *et al*, 2015, p. 2)[1]. Finally, instead of rationality and sober consideration of the costs and benefits assumed by the positivist model, local reaction
to development is often characterised by *emotional* responses of anger and fierce
resistance (Lloyd and Peel, 2007, p. 640; Sturzaker, 2011, p. 557).

The positivist model therefore has benefits in enabling the identification of
different *actors* in planning systems, although its *underlying assumptions* are arguably too simplistic as not accounting for the uneven power that groups have
(Pennington, 2000; Rydin, 1985).

Marxian

Orthodox Marxian Approaches

Marxian approaches view capital accumulation as the underlying driving force
of history and the way that space is configured so that 'the (planning) process is
guided, fundamentally, by the anticipated rate of return to capital' (Short *et al*,
1987, p. 29). History is viewed as a dialectical process between the bourgeoisie and
proletariat, capital and labour, while the configuration of space reflects the imperatives of capital (Castells, 1983; Harvey, 1989, 2013; Soja, 2010).

Marxian theory views the state as upholding, regulating and reproducing capitalism (Fox-Rogers and Murphy, 2014, 2016; Peck and Tickell, 2012). Indeed,
through planning systems managing the land supply, Marxian theory argues that
planners and planning systems create conditions in which capital accumulation
can flourish whilst preventing over-accumulation to reduce the number of crises
which capitalism experiences (Castree, 2008; Prior and Raemaekers, 2007, p. 580).
Nevertheless, planners are generally conceptualised as having limited autonomy
as 'deferential pawns', 'handmaidens' or 'agents of power' as accommodating the
'structural imperatives of capitalism' (Fox-Rogers and Murphy, 2014; Castells,

1983; Kiernan, 1983; Underdown, 1985, p. 277). Consequently, Harvey (1989, p. 3, 2013, p. 115) described the 'geographical landscape' of cities as the 'crowning glory of capitalist development', one which capital fashions in 'its own image' and shapes 'according to distinctively capitalist criteria'.

Nevertheless, this fashioning is a constant and changing process, as Harvey (1982, p. 253,, 1989) also wrote about the 'roving calculus of profit' and the need for it to find a 'spatial fix'. Consequently, during the 1960s/70s, when there were few investment opportunities for capital accumulation in the inner-city or former industrial locations, the market was viewed as 'abandoning' it, resulting in 'capital flight' (Jessop, 2004, pp. 3–4; Lake, 1993, p. 90). Flagship, city-centre developments, like Canary Wharf and Baltimore Harbour, are therefore interpreted as sites of capital accumulation which have been remade and transformed from places of production, dereliction and despair to 'safe' sites of fun, consumption and tourism (the politics of 'bread and circuses') (Harvey, 1989, pp. 7–14).

Peck and Tickell (1996, pp. 595–604) and Harvey (1989, p. 8) argued that entrepreneurial forms of governance or the shift from 'government' to 'governance', exemplified in unelected, commercially dominated quangos in Britain (like Development Corporations), represent attempts by planners to accommodate and attract capital back into cities. Likewise, Smith's 'rent-gap' model (1996, p. 46, 51–74) and arguments surrounding the 'revanchist city', interpret gentrification as an attempt by the market to reassert middle-class control over the inner-city and restore it as an 'attractive' place for investment.

Moving Beyond Orthodox Marxist Approaches: The Clash of Community and Capital

However, the 'rural-urban' fringe has been largely overlooked in the Marxian literature, and the Green Belt stands as *the* prime example that the desires and imperatives of capital, in this case (volume) housebuilders, *do not* always dominate space notwithstanding huge development pressure (Gallent and Shaw, 2007, p. 617; Prior and Raemaekers, 2007). Indeed, the Green Belt directly *restricts* capital accumulation by housebuilders for whom building there would be very profitable as a highly desirable place to live. This shows the inadequacy of the totalising nature of Marxian theory because the Green Belt restricts capital accumulation in England whereas, in the USA and across the Global South, capital flight and the 'circuits of capital' have created 'edge cities' and sprawl (Gallent *et al*, 2006; Marais *et al*, 2020; Scott *et al*, 2013, p. 9). This highlights the critical importance of geographical heterogeneity.

Of course, in Marxian terms, the Green Belt could be read as an example of the state regulating capitalism to maintain capital accumulation. In one sense, housebuilders benefit from the policy because they (are argued to) contribute towards high house prices and maintain profitability whilst preventing the damaging oversupply of housing which often results in dramatic house price reductions and volatility, like in Spain and Ireland during the Financial Crisis (Amati, 2007, p. 591; Kilroy, 2017, p. 6; Rydin, 1985, p. 37). Dear and Scott (1981, p. 13) argued that the policy is a 'historically-specific and socially-necessary response to the self-disorganising tendencies of privatised capitalist social and property relations…in space'. However, many developers are critical of and frustrated with the Green Belt, and affordable housing is essential for homeownership and the continued reproduction of capitalism (Ball *et al*, 2014; Foglesong, 1986).

Drawing on the literature, the role of opposition to development from home-owners is key to the Green Belt's success and longevity (Mace, 2018). Lake (1993, pp. 88–90), drawing on a North American context, described NIMBYism as 'the role of place in the mobilisation and empowerment of community interests against the interests of capital'. He helpfully characterised clashes over housing develop-ments as a conflict between 'capital' and the 'community' (although 'communities' are not homogeneous as divided, in particular, between homeowners and non-homeowners)[2] (Short *et al*, 1987, p. 37; Alexander, 2010).

The rise of mass homeownership has been vital in capitalism's success and longevity as bringing many people *into* the capital accumulation process, with property being the most valuable asset that most people own, with 85% of British household wealth stored in housing (Edwards, 2015, 2016A; Sturzaker and Shucksmith, 2011, p. 174). Although largely unforeseen by Marx, mass home-ownership means that society is more complicated than Marx's dialectic of la-bour and capital, creating conflicts, tension and fragmentation *within* capital or, as Foglesong (1986, p. 22) argued in relation to the Progressive Era in the USA, there can arise 'intra-capitalist conflict' (Cherry, 1982; Foglesong, 1986, p. 22). Drawing on this neo-Marxian theory, homeowners can be conceptualised as of-ten directly conflicting with another group of capital, (volume) housebuilders, with homeowners strongly desiring to protect property value while housebuilders make a profit out of building new houses (Short *et al*, 1987, p. 31). This conflict is played out in the broader arena of planning systems but the Green Belt, as the strongest planning protection against development, is especially fiercely pro-tected by homeowners and equally contested by housebuilders as restricting their capital accumulation.

More broadly, there is the 'homevoter' hypothesis in the North American lit-erature, where homeowners can vote for restrictive zonal ordinances, but people opposing development due to house prices as the primary motive, especially in England with its popular privileging of homeownership, is prominent in the broader literature (labelled the 'house price hypothesis' in this book) (Dehring *et al*, 2008, p. 155; Fischel, 2001A, 2001B). For example, Coelho *et al* (2014, p. 12), in their quantitative study of England's (then) 349 Local Planning Authorities, found that a 10% higher homeownership level correlated with a 1.2% fall in the growth of housing stock between 2001 and 2011.

Short *et al* (1987, pp. 36–37), in a case study of housebuilding in Central Berk-shire and using a neo-Marxian perspective, argued that there was 'a hardcore of material interest underneath the environmental concerns, relating to the impact of new development upon house prices'. Moreover, in a study of the Metropolitan Green Belt in Essex, Rydin (1985, pp. 64–65) argued it served the 'interests of certain sectional groups, particularly already powerful economic interests' who desired to preserve their 'elite situation' and 'high amenity, high-value housing' (see also Simmie's (1981) study of the Metropolitan Green Belt in Croydon). This makes the Green Belt a vitally important object of study as potentially a clear lens through which to view broader struggles in planning systems and an 'arena' for wider societal conflicts, particularly why communities often oppose housing development (Short *et al*, 1987, p. 36).

Whereas Harvey (1989, p. 6) stressed the importance of 'coalitions' of capital, the struggle over new housing development could be better characterised as an intensive 'intra-capital *conflict*' between homeowners and housebuilders so this neo-Marxian framework still takes a largely conflictual view of society (Foglesong, 1986, p. 22; Kiernan, 1983, p. 72). Lake's (1993, p. 89, 90) 'capital' and 'community' framework or, adapting and refining it to the contemporary realities of the rural-urban fringe – the 'homeowners' and 'housebuilders' conflict – is a useful characterisation and this book's overarching theoretical framework to be explored in the empirical chapters.

The Green Belt has not been explored in major academic studies through a neo-Marxian lens. Moreover, apart from Rydin's (1985) cursory acknowledgement, the specific link between the Green Belt and house price hypothesis, which is central to the overarching research question about the housing crisis, has not been empirically explored, notwithstanding the common assumption in the literature that the Green Belt maintains high property prices.

Finally, notwithstanding arguments surrounding Marxian theory not sufficiently accounting for the normative and cultural, arguably 'normative' and 'rationalistic' support for the Green Belt is *intertwined* and difficult to disentangle (Mace, 2018, p. 19)[3] with Matthews *et al* (2015, p. 68) arguing that 'economic capital invested in housing is converted into symbolic capital'.

Other Marxian Approaches and Their Application to the Green Belt

THE CAPITALIST-DEMOCRACY CONTRADICTION

There are other studies which give more autonomy and agency to planners as 'market actors' and acknowledge the complexity of society (Castells, 1983; Heurkens *et al*, 2015, p. 625). For example, Foglesong (1986, pp. 21–24), in writing about city planning in the American Progressive Era (1890s–1920s), argued that planners are the great organisers and mediators of compromises in the 'capitalist-democracy' and 'property' contradictions. The capitalist democracy contradiction is essentially the need for planners to balance and reconcile the imperatives of capital accumulation with guaranteeing that capitalism has (apparent) 'democratic legitimacy' through democratising and socialising space (Foglesong, 1986, p. 23). How far the market determines planning outcomes as opposed to the desires of local communities has been a perennial, fundamental tension in planning systems (Sturzaker, 2017).

The Green Belt places a significant amount of power in the hands of planners to restrict capital accumulation, and although it has limited public access, arguably, the policy being popularly viewed as the countryside and 'commons' means that it has been successfully 'socialised' as space (Bradley, 2019B, p. 695; Evans, 1996; Harrison, 1981). The property contradiction is basically the aforementioned 'intra-capitalist conflict' with Foglesong (1986, p. 22) arguing that, for example, businesses and manufacturers often desire affordable housing for their workforce, which regularly conflicts with the desire of landlords to rent their houses at the highest rate or housebuilders to build the most profitable homes. The theoretical framework of housebuilders and homeowners utilises this helpful conceptual framework of an 'intra-capitalist conflict'.

Regulation Theory

Another fruitful way that the Green Belt has been interpreted is Prior and Raemaeker's (2007, p. 581) application of regulation theory. Marxian theory posits that the governance of cities is flexible and adjustable, especially in a neoliberal economic context (see Peck and Tickell (1996) on Manchester's regeneration and Harvey (1989) on Baltimore). Regulation theory is broader and concerns the governance of whole economic systems as viewed in distinct historical phases or 'regimes of accumulation' in which governance arrangements and the economy are structured to facilitate capital accumulation (Barlow and King, 1992; Castells, 1983; Lake, 1993, p. 88). For example, Fordism, which emerged from the Great Depression and WWII, essentially involved mass manufacturing and consumption, trade unions, the welfare state and a large degree of state control over the economy (Jessop, 2004). Prior and Raemaekers (2007, p. 596) argue that the Green Belt was implemented in this context, alongside new towns and regional policy, as premised on the view that national Government could control and manage economic growth and population change whilst developing most new housing itself. However, while the economy and society have changed vastly with the Fordist system swept away by globalisation/neoliberalism, the Green Belt remains as an 'immovable shibboleth' in the face of the 'irresistible force of the spatial requirements of a post-Fordist economy' and 'shackled to an obsolete Fordist world view' (Prior and Raemaekers, 2007, p. 596). This historical reading makes the context in which the policy was introduced intelligible, although it fails to explain *why* it has remained as a policy. The rational and normative aspects of its popular support are therefore key in explaining its longevity (Mace, 2018, p. 15).

Marxian Language

Objectors to development often use the language and tone of Marxian language despite not being Marxists themselves, as Warren and Clifford (2005, p. 366) found regarding the St Andrews Green Belt (see Chapter 7). Firstly, there is the 'rational' side of seeking to prevent development by trying to discredit (volume) housebuilders as the 'other', 'greedy' and wanting to 'just to make a profit' at the expense of the local area compared to CPRE's 'pure' image (Sturzaker, 2011, p. 556, 564; Sturzaker and Shucksmith, 2011, p. 175). In this way, objectors seek to generate moral outrage to mobilise local opposition even though, morally, a developer making profit from housebuilding is no different to supermarkets making a profit on the most basic necessity – food. Secondly, there are 'normative' arguments that there should be things, like the 'countryside' and 'environment', which are more important than profit and should be protected by planning systems, which is an interesting simplification of the Marxian argument about society becoming increasingly commodified (Harvey, 2009, 2013).

Poststructuralist Analysis and Collaborative Planning

Marxian theory has been challenged by poststructuralist approaches, which generally argue that the dualism of labour and capital is too simplistic to explain the complexities of contemporary society, which is marked by difference and wider divisions based on gender, ethnicity, age, etc. (Allmendinger, 2009, pp. 156–157, 226). Poststructuralism often relates to the concept of pluralism about the multiplicity of interest groups in society and the complexity of planning systems (Fox-Rogers and Murphy, 2016; Sandercock, 1998, p. 75). The salience of the cultural and emotional in campaigner opposition to development has been underlined, especially the importance of place attachment, as highlighted in Davison *et al's* work (2016, 2017) on affordable housing in Australian cities and the broader environmental psychology literature (Anton and Lawrence, 2014). The poignancy of these arguments can been seen in the Green Belt's emotional importance (see Chapter 3) and the limited effectiveness of compensation in increasing homeowner support for housebuilding (Inch *et al*, 2020; Ministry of Housing, Communities and Local Government (MHCLG), 2018). Additionally, research has found that financial payment to local residents has been ineffective in changing attitudes towards windfarms internationally due to the importance of people's principled opposition (see Devine-Wright's, 2009 and others work – for example Cowell *et al*, 2011; Gross, 2007).

A response to the diversity and difference in contemporary society is collaborative planning on which there is a vast literature (for example, Brand and Gaffikin, 2007; Innes and Booher, 2015, p. 284). Collaborative planning emerged from the concept of 'deliberative' democracy and 'communicative rationality' as it is argued that the role of the planner is to listen to people's views and seek to arrive at consensus among competing groups (Healey, 1997, p. 16, 2003, p. 104).

However, poststructuralism is arguably limited in how far it accounts for the 'embedded' and uneven power relations in planning systems (Healey, 1997, p. 59; Tewdwr-Jones and Allmendinger, 1998). Consequently, this book's theoretical frame accepts the poststructuralist view that there are a multiplicity of interests in planning but largely focuses on housebuilders and homeowners who the literature identifies as having the *most* power (Rydin, 1985, p. 64).

Agonist Planning

Agonist planning theory is a popular theory which accepts the plurality and multiplicity of groups in society but advocates a more conflictual or adversarial response of groups working to gain concessions from the 'system' (Chettiparamb, 2016, p. 1286; Lennon, 2017, p. 154). The theory is usually associated with Mouffe (2000, p. 427) and was introduced into planning theory by authors such as Hillier (2003) and Pløger (2004). Although premised on adversarial engagement, it aims to secure positive gains for groups involved so is different to 'antagonism' (McClymont, 2011; Vigar *et al*, 2017, p. 426). Agonist theory has been applied helpfully to Neighbourhood Planning, especially the desires of some neighbourhood groups to pursue more radical and innovative policies (such as banning second homeownership of new homes built in St Ives and Mevagissey (Cornwall)) (Bradley, 2018; Wargent and Parker, 2018, p. 384). Studies have also shown how debates have often been 'closed down', 'modulated' or circumscribed by Neighbourhood Planning groups in order to forge consensus and complete Plans or by 'absent others', typically civil servants and the Planning Inspectorate, placing tight limits and restrictions on what they can achieve (Parker and Street, 2015, p. 794; Parker *et al*, 2015, p. 519).

Nevertheless, its utility when applied to the Green Belt is more limited. As argued in Chapter 8, Green Belt campaigners regularly adopt an agonist approach in the way that they fiercely oppose attempts by LPAs to change Green Belt boundaries through protests, leaflet drops and social media campaigns (Sims and Bossetti, 2016). However, alongside 'working against' councils and developers in an agonist way, they also regularly try to 'work with' the system by adopting more consensual, formal approaches, such as lobbying MPs, the Secretary of State and councillors and employing 'planning language' (Amati, 2007, p. 285, 291). Indeed, this highlights how campaigners often play a 'mixed game' of consensus and confrontation (Wargent and Parker, 2018, p. 393). Nevertheless, groups benefiting from more affordable housing often form a too geographically dispersed group to mount effective agonist engagement alongside lacking the necessary time, money and resources (Myers, 2017). This illustrates Barry *et al*'s (2018, p. 420) and Inch (2015, p. 418) critique of agonist planning being limited in its effectiveness because of the 'inequality of arms' in resources between different groups.

Towards an Analytical Framework: The Spatial Justice and Just City Approach

This section empirically operationalises and explores this study's *theoretical* framework through developing an *analytical* framework, based on the concept of social justice and the just city, which helps to develop policy recommendations and shape the methodology (Chapter 2) (Fainstein, 2010, 2014). Again, this is vitally important in the current context of the deepening housing crisis and the Green Belt being widely charged as a significant cause of it (Sturzaker, 2017).

The main advocate of the just city approach, Fainstein (2010, 2014), began with a similar starting point to this book of being disillusioned with the limited utility of Marxian *and* collaborative planning approaches in explaining contemporary realities and offering a workable way forward in policy terms[4]. Fainstein (2010,

p. 5, 2014, p. 7) contended that just, collaborative processes, important as they are, are insufficient *in themselves* to guarantee just *outcomes*. Consequently, Fainstein (2010, p. 5, 2014, p. 7) developed the *just city* concept with diversity, democracy and equity as its key criteria and used these to evaluate Amsterdam, New York and London. The just city is therefore a workable framework to provide practical measures to improve capitalism and serve the public interest through delivering social justice (Fainstein, 2010, p. lx,, 2014, p. 14).

Soja (2010, p. 75, 92) also accused Marxism of not offering scholars advice 'short of the total transformation of capitalism' and developed the concept of *spatial justice* or 'equitable access to urban resources'. More recently, Lake (2016, p. 1208) made an appeal for social justice not just to be the 'object' of planning systems but also its 'subject', thus aiming to take planning back to its moral, social roots. These lines of thinking can be seen in the influential Raynsford Review (2018B, pp. ix–xi) and wider work of the TCPA (for example, Ellis, 2020; Ellis and Henderson, 2014), which has called for planning to return to a social purpose.

Nevertheless, social justice as an analytical framework in planning research goes back to Hall's work, *Containment* (1973A, p. 433, 1974, p. 406), where he assessed which social groups 'won' and 'lost' from post-war planning systems, especially the Green Belt. Likewise, the North American academic Kiernan (1983, p. 72, 81–83) launched an attack on the rational/unitary public interest model, which argued that planning can 'reconcile and satisfy all…interests'. He developed a conceptualisation of planning being primarily conflictual as composed of 'competing', not 'congruous' groups and highly political, although, like Soja and Fainstein, he also powerfully attacked Marxian theory[5]. Nonetheless, reflecting Hall's (1974) powerful critique of the Green Belt, Kiernan (1983, p. 74) argued that in planning, 'some benefit and some lose…the same groups usually benefit and other groups usually lose' with these effects falling 'disproportionately upon different socio-economic classes'. Kiernan (1983, p. 75) therefore made a powerful appeal for planning systems to return to their 'socially redistributive function' and for the 'primacy of social justice'. This social justice would not involve 'punishment' of the rich but ensuring that there were 'palpable gains' for disadvantaged groups consequent on development (Kiernan, 1983, p. 84). Kiernan's highly practical and pragmatic arguments have resonance with this book's emphasis on practical and realistic policy solutions regarding Green Belt reform.

Other scholars have developed social justice as an analytical framework. Sandercock (1998, pp. 196–198) argued that there are 'overall winners and losers' in planning and, in an evaluation of the redevelopment of Melbourne Riverside (Sandercock and Dovey, 2002, p. 152), she argued that researchers/planners should ask:

> Who gains and who loses in this city-in-the-making and by which practices of power? Are these outcomes desirable and to whom?

This book asks similar questions of the Green Belt. Flyvbjerg (1998) has explored uneven power relations in planning systems and argues for 'phronetic planning research', which has the study of power at its centre and research which is pragmatic, practical and applied (Flyvbjerg, 2004, pp. 284–293). Flyvbjerg (2004, p. 290) argued that research should ask:

1 *'Where are we going?*
2 *Who gains and who loses, and by which mechanisms of power?*
3 *Is this development desirable?*
4 *What, if anything, should we do about it?'*

Again, this book asks these questions in this analytical framework and the methodology (Chapter 2) with the assessment of whether the planning system and the Green Belt serve the 'public interest', as evaluated through social justice, being its principal focus, especially as Satsangi and Dunmore (2003) argued that planning has focused too much on conservation compared to social sustainability.

Finally, the comprehensive philosophical literature on different types of justice, especially associated with Rawls, is acknowledged, although space does not permit a detailed examination here (Chettiparamb, 2016, p. 1288; Lennon, 2017, p. 151). For example, there is both intergenerational and intragenerational justice (Sturzaker and Shucksmith, 2011, p. 175). Likewise, Low (2013, p. 295) helpfully highlighted that there are three aspects to justice: procedural, distributional and interactional. This book, drawing on Fainstein's (2014, p. 7) distinction, aims to examine both planning *processes* and outcomes, especially intergenerational and intragenerational justice. Consequently, whilst not *directly* employing Fainstein's (2010, p. 5) framework of diversity, democracy and equity, it draws on her overarching *theory* of social justice to assess 'who benefits and who loses' from the Green Belt and wider planning system.

Developing the Just City Framework Further: The Feasibility of Green Belt Reform

Chapters (3/4) have demonstrated the Green Belt's normative and rationalistic popularity and longevity, whilst the social justice approach advocates workable policy recommendations (Fainstein, 2010, 2014). Indeed, policy recommendations which do not sufficiently acknowledge the Green Belt's social and political significance, such as advocating its abolition, are likely to fail from a social justice perspective, yet successfully solving the severe housing crisis is still a vital current priority (Mace, 2018).

The 'feasibility test' will be applied to potential policy recommendations (Table 5.2) (Breheny, 1997, p. 209). This test was developed by Breheny (1997, p. 210) when critically evaluating urban compaction and brownfield-first policies, but the premise of Breheny's argument, that planning policies need to be politically and socially acceptable in order to be workable (as reflected in the broader policy evaluation literature (for example, Marsh and McConnell, 2010; McConnell, 2010; Palfrey *et al*, 2012)), is a valuable insight. Apart from Mace *et al*'s (2016) innovative study on the Metropolitan Green Belt, arguably most Green Belt studies,

Table 5.2 The feasibility of Green Belt reform

Practical/
  Environmental
Housing Crisis
  Chapter (6)
	Whether sufficient land exists in the Green Belt, without environmental protections/ restrictions, to meet housing demand at acceptable market densities. The approach taken in most studies, like Cheshire (2014) and AECOM (2015), on the Metropolitan Green Belt.

Economic *Housing Crisis* *Chapter* (6)	*Number:* Whether housebuilders can build enough homes in the Green Belt to meet demand and stabilise house prices with sufficient levels of profit (Archer and Cole, 2014). *Type:* 'Right' type of homes that people 'need', e.g. affordable homes, bungalows or 2–3 bedroom houses, rather than the most profitable (e.g. one bedroom studio flats or executive housing) (Jefferys *et al*, 2015). Is there scope to diversify the market with affordable, self/custom build, social and rental tenures of housing (Archer and Cole, 2014)? *Speed of Delivery:* Can new homes be delivered at sufficient speed to stabilise house prices and meet demand (DCLG, 2017; Rydin, 2013)?
Social *Community* *Chapter* (7)/ Political *Politics and* *Governance* *Chapters* (8/9)	Crucially, this involves whether communities will accept Green Belt reform alongside politicans locally and nationally, given planning's political nature (Munton, 1986). Are there occasions when Green Belt development has been politically and socially acceptable and what factors lead to some Green Belt campaigns being more effective? Housebuilding will arguably be never extremely popular with local homeowners but are there ways in which Green Belt development can be more politically and socially *acceptable*?

especially those advocating TOD there (for example, Clarke *et al*, 2014), do not acknowledge social and political feasibility sufficiently by focusing on practical or economic feasibility. The Green Belt's history demonstrates that reforms cannot be 'forced' through, and social and political opposition will not somehow 'fall away' in the future (see Chapter 3). This book therefore seeks to fill this significant research gap through developing a broader feasibility framework of housing development which is cognisant of social and political feasibility alongside the still important economic and practical feasibility, thereby further developing Breheny's (1997, p. 209) test. These form the structure of the empirical chapters.

Power and Politics in the Planning System

Moving from *which* interest groups have the most power to *how* competing groups seek to influence and exercise power in planning as a 'system of negotiation', the modernist view of planning that planners had a comprehensive understanding of the public interest and were apolitical, was quickly discredited (Barlow and King, 1992, p. 397; Faludi, 1973). Indeed, the amount of commercial influence in post-war redevelopment soon became apparent in the post-war era, especially in Newcastle upon Tyne and with the Ronan Point disaster, which showed that planning sometimes served other (commercial) interests as susceptible to lobbying (McLoughlin, 1985; Pendlebury, 2001; see also Altshuler's (1966) work in the USA).

Planning has always arguably been political because, at its heart, it is about managing a scarce resource, land, over which there are invariably competing interests, thereby necessitating mediation and compromise (Alexander, 2010; Campbell *et al*, 2014; Lloyd, 2006). Planning is deeply involved with people and their everyday lives both directly and indirectly (Warren and Clifford, 2005, p. 355). The literature has been highlighted on how planning *outcomes* regularly reflect those who have *power* in planning, especially homeowners and developers, although planning is nominally democratic (Fox-Rogers and Murphy, 2014, p. 263). As Cherry (1982, p. 116) argued: 'planning reflects the interests of those who wield power and influence...there have been some obvious beneficiaries (the middle and upper suburbanite or the land or property speculator)'.

The *way* in which homeowners seek to influence planning, how power operates and its political nature is explored through Lukes' (2004, pp. 14–59) *three dimensions of power* and Sims and Bossetti's (2016, p. 35) and Fox-Rogers and Murphy's (2014, p. 250) framework of formal and informal political power.

Luke's Three Dimensions of Power

This helpful framework, which seeks to explain how a 'dominant group' exercises power, is primarily used here alongside other frameworks, such as Fox-Rogers and Murphy's (2014, p. 244) 'informal strategies of power', to explore the often more subtle or informal ways that power operates (Lukes, 2004, p. 28; see also Sturzaker and Shucksmith (2011, p. 170))[6]. It is applied to the Green Belt as the literature suggests that homeowners and housebuilders are the dominant groups at the rural-urban fringe and that planning outcomes often reflect their desires (Table 5.3) (Hall, 1974; Sturzaker, 2010). Lukes' framework, which seeks to explain how a 'dominant group' exercises power, is primarily used in this chapter to examine the *modalities* of power, especially the more subtle or informal mechanisms (Lukes, 2004, p. 28; Sturzaker and Shucksmith, 2011, p. 170).

As explored later in Chapter 8, Lukes' framework has been widely critiqued including for not fully addressing the philosophical question of what power 'is', especially the difficulties of establishing causal links between the dimensions and their impact on policy alongside challenges in distinguishing between the decision and non-decision-making process (Dowding, 2006; Robinson, 2006). Reflecting on these critiques, Lukes (2004, p. 150) highlighted the importance of empirical studies in relation to the dimensions and stressed that assessments of power are necessarily 'partial and limited'. Additionally, the framework has been critiqued as outdated, but it has been revised in the light of Foucauldian critiques and arguably still has relevance and utility in terms of elucidating different modalities in the exercise of power (Plaw, 2007). Indeed, Haugaard (2021, p. 153) has recently added a fourth dimension relating to the 'social construction of subjects in response to context'

Table 5.3 Lukes's three dimensions of power

Dimension	How power is exercised	Evidence
First Dimension: *Who makes decisions.* Most basic level of power with the *outcome* of decision-making reflecting the most powerful (Lukes, 2004, p. 16). The 'overt' exercise of power (Rydin, 1985, p. 58).	Decisions about local plans/major applications made by elected councillors in planning committees. Although nominally representing the *whole* community, certain groups, especially the 'powerful anti-development lobby', are vocal/influential (Sturzaker, 2010, p. 1007). Planning Committees largely reflect the *demography* of homeowners being unsalaried: white, middle-class and older. Political interests sometimes determine decisions.	Guildford Local Plan: Green Belt release scaled down during an angry consultation process with 32,000 responses (Edwards, 2016B). The plan was subject to a (failed) judicial review. Sims and Bossetti (2016, p. 36) found that 40% of councillors in outer London Boroughs said being supportive of more housebuilding loses votes. Greater Manchester Spatial Framework: Green Belt release reduced by 60% after 44,000 responses/ widespread protest.
Second Dimension: *The 'decision-making and non-decision-making' process* (Lukes, 2004, p. 22). 'Covert' exercise of power (Rydin, 1985, p. 58).	Consultation processes for Local Plans/applications are dominated by the loudest voices, with other voices largely 'hidden' (Lloyd, 2006, p. 10). Vocal opposition to housing can intimidate voices from speaking in support through fear of social ostracisation, illustrating Bourdieu's (2005, p. 92) concept of 'symbolic violence' (Cited in Sturzaker and Shucksmith, 2011, p. 182).	Wisley Airfield (MGB, Surrey). Application for 2000 homes refused but the developer found that 2/3 of local young people supported the proposal despite well-organised protest (Edwards, 2016B). Amati (2007, p. 585) found that most objectors to housing in the MGB are 'experienced' (previous objectors who can effectively use planning 'language').
Third dimension: *The dominant group's discourse. Decisions are accepted by people as the 'existing/natural' order of things'* (Bourdieu, 2001, p. 35 cited in Lukes, 2004, p. 28). 'Latent', 'most effective, insidious' exercise of power (Rydin, 1985, p. 65). This involves the powerful 'influencing, shaping or determining' people's 'wants' and, controversially, 'suppressing [their] unrealised interests' (Lukes, 2004, p. 27).	Power of discourse/imagination in idealising Green Belts while opposition to housebuilding is galvanised by the dominant myth of the countryside being 'concreted over'. The Green Belt is viewed as 'sacrosanct', 'precious' and 'natural'. This discourse is so powerful that many city dwellers and renters support it in what Sturzaker and Shucksmith (2011, p. 189) label 'thick acquiescence/symbolic violence' as excluding people from living in the countryside.	This discourse is often employed by CPRE/local campaigns. Many people confuse greenfield and Green Belt sites and assume that Green Belt boundaries can never be altered. CPRE often dominates housing debates, with Pennington (2000, p. 195) finding that in *The Times*, CPRE was five times more referenced than competing interest groups. Cheshire (2013, p. 1) argues that Green Belts are (viewed as) as keeping 'unwashed urbanites corralled in their cities'. Mass media is used as an example of this 'process of socialisation' (Plaw, 2007, p. 489).

From Maidment (2016), Morphet (2011, p. 130), Sims and Bossetti (2016, pp. 9-15) and Yarwood (2002, pp. 277–282),.

Politics and Politics

Sims and Bossetti's (2016, p. 35) framework for highlighting how politics, power and interest groups operate in planning is applied here to explore planning's political nature. Sims and Bossetti (2016, p. 35) usefully highlighted that the 'Politics', formal political process, and 'politics' of planning, more informal but the wider political culture, including campaigns, the media and conversations, conjoin and determine planning outcomes. However, space is vitally important in how 'Politics' and 'politics' operate nationally and locally, especially as the national Government is heavily involved in planning with England having a highly centralised state (Peck and Tickell, 1996, p. 612; Short *et al*, 1987, p. 39). Likewise, Fox-Rogers and Murphy (2014, p. 244) have highlighted the importance of 'informal strategies of power' in the Irish planning system, so these strategies are explored here (Table 5.4) and empirically elucidated in Chapter 8.

Table 5.4 'Politics' and 'politics' in planning

Interrelated	*'Politics'*	*'politics'*
↑ ↓	National: • National planning: NPPF, Green Belt policy, Population Projections (Standard Method), Housing Delivery Test (HDT). • Secretary of State call 'in powers' for appeal/local plans (Birmingham/Bradford) • Electoral campaigns: promises by Party Leaders and Manifestos pledges to protect Green Belt, esp. in 'marginal' seats • Lobbying by MPs/Green Belt issues raised (e.g. Andrew Mitchell/Paul Beresford). Local: • Inherently adversarial application process often marked by 'Decide-Announce-Defend' (DAD). Approval of local plans/applications by elected councillors and Green Belt issues in elections (2019). • Local plan examination process/appeals system run by the Planning Inspectorate.	National: • Lobbying/'behind closed doors' access by think tanks, HBF and CPRE. • Political: Pressure by constituency associations and party membership • Media: Influential Green Belt/campaigns, e.g. *The Telegraph's* 'Hands off our land'. • National campaigns by CPRE and other organisations (National Trust, Shelter and YIMBY movement). Local: • Consultation process and Green Belt campaigns (in Sutton Coldfield (Project Fields)), St Albans and Guildford[7]). Social media, leaflet drops and media campaigns to influence politicians. • Lobbying of councillors via technical 'kinship' (Amati, 2008, p. 390).

From Boddy and Hickman (2018), Carpenter (2016), DCLG (2012), Edgar (2015, p. 1), Edwards (2016B), Morrison (2010, p. 159), Rydin (2011, p. 95), Sturzaker (2017) and Walker (2016, p. 208).

Conclusions from Power and Interest Groups in Planning Chapter

The aims of this chapter have been fourfold. Firstly, to explore planning theories to evaluate how far they reflect the realities of the Green Belt. Secondly, based on these theories, to build an overarching theoretical framework. Thirdly, based on this theoretical framework, to develop an analytical framework and shape the research questions. Fourthly, to elucidate the theoretical and analytical framework by exploring the ways in which power is exercised in planning systems. This chapter has therefore aimed to theoretically conceptualise Chapters 3 and 4 by viewing the Green Belt as the arena in which wider planning and societal conflicts are played out. It provides a link between these history and evaluation Chapters (3 and 4) and the subsequent empirical Chapters (6–9) by conceptually framing them

This chapter has shown the limitations of an orthodox Marxian approach by arguing that planning systems and society are more complicated than a simple struggle between capital and labour, especially regarding the Green Belt (Cherry, 1982). Post-structuralist approaches are also limited as not sufficiently accounting for the uneven power relations in planning systems, particularly the power of homeowners *and* housebuilders (Lake, 1993; Sturzaker, 2010). Consequently, both *overall* Marxian and post-structuralist theoretical frameworks are rejected, with this chapter conceptualising the Green Belt as an 'intra-capitalist conflict' between homeowners and housebuilders (Foglesong, 1986, p. 22; Lake, 1993). Building on this framework, it has argued that social justice as an analytical framework, especially evaluating 'who benefits and who loses' from the policy, is a useful way to operationalise and elucidate this broader intra-capitalist conflict framework (Hall *et al*, 1973A; Kiernan, 1983, p. 83). Moreover, this social justice approach attempts to offer practical reforms to capitalism, hence providing a useful 'bridge' between theoretical conceptualisation and practical policy recommendations for the Green Belt (Soja, 2010). The section on the *way* in which power is exercised in planning elucidated the theoretical and analytical framework and developed the context for constructing the research questions.

Notes

1 It is acknowledged that 'development', especially the current volume housebuilder dominated model, does not invariably benefit non-homeowners due to the cost of new homes but this does not negate the point regarding the power of campaigners.
2 Those benefitting from affordable housing, like renters, children and future generations, often do not get involved in this conflict as a geographically and socially dispersed group (Whitehead *et al.*, 2015, p. 4).
3 As explored in Chapter 7, 'rationalistic' support of the policy relates to house prices and the material impacts of development for individuals where 'normative' relates to more principled support (Mace, 2018, p. 4).
4 Fainstein (2010, p. lx, 2014, p. 14) argued that Marxism was flawed because its main aim, to abolish Capitalism, is largely unrealistic and undesirable, especially with the failure of successful or viable alternatives.
5 For example, Kiernan (1983, p. 81) argued that Marxism is 'stronger in diagnosis than prescription' and that its 'sweeping nihilistic analysis...scarcely leaves much room for

the application of public policy' as demanding nothing less than the 'total systematic transformation of the capitalist order'.

6 The limitations of Lukes' work, especially the more theoretical question of what 'power' is and empirical question of establishing causal links, are acknowledged and explored in Chapter 8.

7 In Guildford, two anti-development political parties were formed: the Guildford Green-belt Group (GGG) and Residents for Guildford and Villages (R4GV). GGG won four seats and nearly 7,000 votes whilst R4GV won 15 seats and nearly 20,500 votes in the 2019 Local Elections (Brock, 2018).

6 The Green Belt, the housing crisis and policy reform

Introduction

This chapter explores the book's central theme, the (housing) crisis, and is policy-orientated as well as theory-related as laying the empirical groundwork for contextualising wider issues, including governance, politics, and community opposition, examined later in the book. It directly addresses the overarching research question through evaluating the extent of the Green Belt's responsibility for the housing crisis, examining how far it needs to be reformed and reflecting upon the implications for other Urban Growth Boundaries internationally.

As Chapter 4 demonstrated, economic studies in the UK and elsewhere regularly argue that planning restrictions are *a* cause of housing crises in many countries, yet they often struggle to *disentangle* the *extent to which planning policy is specifically responsible for house price increases, especially in relation to the Green Belt* (Hilber and Vermeulen, 2014; Kenny *et al*, 2018). Moreover, the *discursive* or *socially constructed* nature of housing is increasingly being recognised by studies with the 'linguistic turn' in housing studies, so it is important to consider how planners/campaigners view and construct the housing crisis through their own worldviews to triangulate and view it from different angles (Harrison and Clifford, 2016; Munro, 2018, p. 1092). Indeed, as the key actor in the planning system, how planners view the housing crisis is vital to policy and practice, especially for developing recommendations grounded in what is politically possible rather than just theoretically desirable (Campbell *et al*, 2014; Kenny, 2019B).

This is crucial given the severity of the housing crisis and its political importance, yet the Green Belt is still probably England's most long-standing and popular planning policy (Dockerill and Sturzaker, 2019). The chapter also relates to planning theory by empirically examining the neo-Marxian conceptual framework. The first half of the chapter deals with causation, and the second half with recommendations. Although focusing on *policy*, the chapter sits within the heart of the book's broader recommendations surrounding the need for longer-term, strategic governance and a national Green Belt debate.

DOI: 10.4324/9781032674315-6

To What Extent Is the Green Belt Responsible for the Housing Crisis?

The Housing Crisis: A Disputed Term

General Context

The term 'housing crisis' is a widely-used, if controversial, one (Gallent, 2019; Harris, 2019). The main reason for this book using the term 'housing crisis' is because it is the most widely used, popular term to describe England's housing 'problem', thereby linking the book to contemporary political and policy agendas, especially with some now calling the housing crisis a housing 'emergency' (Airey and Doughty, 2020; Young, 2023, p. 1). Additionally, the phrase links this study to the broader concept of shared housing crises around the world rather than focusing solely on the UK (Christophers, 2018, 2019).

The 'crisis' is primarily an *affordability* one whereby the high level of house prices and their rising nature makes it hard, particularly for young people, to cobble enough money together for a mortgage deposit, thereby 'locking' them out of homeownership and forming households and families (Christophers, 2019; Gallent, 2019, p. 489; Ryan-Collins, 2018, 2021, p. 480). Moreover, privately rented housing is typically more expensive, insecure and poor quality than owner occupation. As a West Midlands regional policy expert argued:

Young, newly forming households cannot get decent accommodation at a reasonable price either to buy or to rent, that is my definition of the 'housing crisis'…Everyone should be able to get a house at the price they can afford.

The term 'crisis' is critiqued because there have always been housing problems in Britain (and many countries) and people who are poorly housed (Gallent, 2019; Lund, 2017, 2019). However, affordability and housing issues have intensified greatly recently as, until the 1980s, a large amount of social housing was built, providing an alternative form of accommodation whilst, until the mid-2000s, access to mortgages for homeownership was relatively affordable to most people (Bramley, 2019; Bramley *et al*, 2017; Griffith and Jefferys, 2013). Consequently, the growth of house prices and privately rented accommodation conjoining with the lack of social housing have caused a problem which arguably merits the term 'crisis' or what some, even on the Right, have called a 'catastrophe' (Rees-Mogg and Tylecote, 2019, p. 11). Housing costs are the biggest driver of intergenerational and intragenerational inequality in England, and this inequality is reaching Victorian levels whereby one's ability to purchase a house and wider life chances depend more upon inherited wealth or the 'bank of Mum and Dad' than one's ability and earning potential (Halligan, 2019, p. 3).

There are other societal issues related to the housing crisis including the geographical immobility of labour, housing insecurity for future retirees, the lack of suitable 'family' homes and homes for the elderly to 'downsize' to, dependence upon volume housebuilders for most housebuilding, the cost of land, second homeownership, rising homelessness and the 'overcrowding' of many homes compared to the presence of spare bedrooms in others (Dorling, 2015; Hudson and Green, 2017; Kenny, 2019A; Morphet and Clifford, 2019). Indeed, the complex, multifaceted nature of the housing crisis merits the term *housing* crisis rather than just *affordability crisis* and means that there are no immediately straightforward solutions, although this study focuses on owner-occupation (Kilroy, 2017; Pendlebury, 2015; RTPI, 2016A). Moreover, the dependence of the British economy upon rising house prices means that 'radical' solutions designed to 'crash' the market would be dangerous, as Planning Academic (2) warned. This clearly justifies this book's approach of exploring realistic recommendations regarding housing rather than anticipating a transformation of the mode of production (Fainstein, 2010, 2014; Monbiot *et al*, 2019). Conversely, the severity of the housing problem as potentially posing an existential threat to the current economic system provides a further basis for labelling it a 'housing crisis' (Donnelly, 2019). In this regard, the word 'crisis' is used politically as a problem which urgently requires solutions (Young, 2023).

Another critique of the word 'crisis' is that there are multiple 'crises' with the housing market being marked by spatial 'nuance' and heterogeneity and that the 'crisis' often written about in the media is more accurately the housing crisis in London or other world cities (Kilroy, 2017; McKee *et al*, 2017, p. 60). However, whilst the housing crisis is *most severe* in the Greater South East (hence the study's focus on Green Belts in *regional* England to explore spatial heterogeneity), a similar set of housing issues emerged in the data around England, especially house prices out of the reach of local people (Gallent, 2019; Gallent *et al*, 2019). These shared spatial characteristics merit the term 'crisis' in a geographical as well as a temporal sense.

How Planners and Campaigners Defined the 'Housing Crisis'

Most planners and campaigners acknowledged that there was a serious housing problem reflecting popular discourse, people's lived experiences and how the housing crisis is now firmly rooted in national consciousness (Harris, 2019; Inch and Shepherd, 2019). However, campaigners varied in how seriously they viewed it, with a few comments made by everyday campaigners disputing its existence, with an SSGB campaigner saying, '*the so-called housing crisis is man-made*'. Nevertheless, an example of a typical qualified response by a professional campaigner was:

> You used the phrase 'housing crisis'…there is a danger in that phrase in that it is a little bit too glib. **The way I see it is that there are shortages of a particular types of housing in particular areas. There is no one single national housing crisis.**
>
> (Planner (1), CPRE West Midlands)

This was echoed by other campaigners and retired planners, especially regarding the housing market in the West Midlands:

> Of course there are hot spots, but you can still buy a house...in Bearwood which is 10-15 minutes from Birmingham city centre on a bus...for £100,000, maybe less than that possibly... **The big issue is really about the scale of social housing. Forget the house prices. The house price thing – it's not a red herring but it's over inflated by the development interests.**
>
> (Retired Strategic Planner (2), West Midlands)

However, practising planners invariably referred to the 'housing crisis', and it was often used as an unquestioned, self-explanatory phrase as the main justification for new housebuilding and a more relaxed Green Belt rather than a phrase to be explained and qualified (Bradley, 2023; Gallent, 2019; Inch, 2018). This was particularly pronounced among middle-aged and younger planners who used it to fashion their professional identity and could personally relate to it:

> **At this rate, we are not going to solve the housing crisis through retaining the Green Belt.** So, actually, could there be an opportunity in the future when things get worse that someone might go – 'shall we try it?' [reforming the Green Belt]...Lots of our people in our team pay more to rent than they would if they had a mortgage but they can't get on the housing ladder because they don't meet the criteria or they don't have the £25 grand deposit to buy a house. Sorry, a bit of a sore topic.
>
> (Planners in their 20s at Focus Group II)

Some of the quotes reflected the literature critiquing the notion of the 'housing crisis' internationally, including its geographical intensity in global cities such as New York and London (Inch and Shepherd, 2019; Wetzstein, 2017). However, consensus crystallised around the (un)affordability of housing, so clearly, this should be a key factor in developing policy recommendations in the book. Hudson and Green (2017) convincingly argued that it is vital to consider what 'success' involves in solving the housing crisis whilst organisations, such as the TCPA (Raynsford Review, 2018A; Ellis, 2020), have articulated a broad vision of housing that is affordable, safe and decent.

These wider qualities are vital, but this book focuses on affordability, including renters *and* owner-occupiers, so 'successfully' resolving the housing crisis would mean that renters, social and private and future/existing owner-occupiers can afford and have access to the tenure of their choice (Griffith and Jefferys, 2013; Jefferys *et al*, 2015). Although inevitably an ideal aspiration, it is important to have an objective when analysing policy, especially housing policy (Hudson and Green, 2017), as outlined in the policy analysis literature (e.g. Marsh and McConnell, 2010; Palfrey *et al*, 2012).

Critically Evaluating the Links Between the Green Belt and the Housing Crisis

As the housing crisis is such a multifaceted problem, even the most trenchant critics interviewed admitted that it would be unfair to *entirely* blame the Green Belt or planning regulation generally for the housing crisis. Indeed, planning is limited in how much it can achieve as only allocating *locations* for development and having limited oversight over housing tenure (Goode, 2023; Lord and Tewdwr-Jones, 2018). Successfully resolving the housing crisis arguably requires a *range* of policy responses across local, regional and national Government departments. Arguably, planning is well-placed to solve the housing crisis given planners' ability and that of the broader system to bring together a diverse range of partners to work on complex solutions, with there being scope for more ambitious planning as there is in the Netherlands and Sweden, for example (Hills, 2020; Whitehead *et al*, 2015; Kenny, 2019A).

Although many planners struggled to identify the *extent* to which the policy is responsible for the housing crisis, those who were most critical highlighted it as being one of the most important causes, especially in the Greater South East. Perhaps the strongest remark was an attack on *Branchout*[1] by a retired, local authority planner who labelled it the 'woe of the West of England', the 'greatest indictment' of planning and called for its removal (Baker, 2018, p. 22). Another strong view was expressed by this private sector planner in *Tripwire* entitled, 'Yes, that's right: I hate Green Belt', which read (Griffiths, 2017, pp. 3–5):

> Green Belt…continues to raise its head as the leading issue in the supply of new housing. I intensely dislike how it is used, appropriated and manipulated by those who often have a lot to lose if it disappears…I know how restrictive and unfair it can be…it can fundamentally create unsustainable patterns of development.

These quotes demonstrate the anger and passion that the Green Belt can generate alongside the depth of frustration that some planners have towards it, although they are not representative of the whole profession. Indeed, even the most critical planners interviewed acknowledged:

> **The Green Belt is, if you like, just one factor which has led to the housing crisis**…It is a very restrictive, inflexible policy and has contributed to the housing crisis in so far as it surrounds most of our major cities…it is most probably more responsible, you know, than simply the countryside policies… because, if you look at where the heat is coming from, it is your London's, Birmingham's and Manchester's to some extent. And therefore that is causing price inflation, you know, the shortage of supply and right type of location
>
> (National Private Sector Planning Director (3))

Interestingly, the language and reasoning of economics featured strongly with the Green Belt being blamed for restricting housing supply and raising house prices, with some even citing Paul Cheshire (a leading economist who critiques Green Belt):

> There has to be some sort of direct link between being in the Green Belt and having that proportion of added value onto a house because of the kind of the security and knowledge that nothing is going to come to you in the future...a considerable value is added on to a house in the Green Belt because of the... designation.
>
> (Local Authority Planner, South East)

> **In any supply and demand equation, if demand is high but you re-strict the supply, the price will go up. So, yes, that can happen in Green Belt**...economically that is just inevitable...in any market, there are win-ners and losers...there are consequences of Green Belt policy, inevitably, for those people who are required to live outside the Green Belt because they can't afford to live in it (you have increased journey times, cost)... The older generation...already living in the Green Belt...they have the best of both worlds in that they are closer to facilities, I guess, and they benefit more from house price inflation (but that is not unique to Green Belt)... **most of the value in property is owned by older people and the ones who are having a problem getting on to the housing ladder are those who tend to be younger**
>
> (Private Sector Planning Director (South East) (2))

This last quote is helpful in drawing out who 'gains' and 'losses' from the policy and the inequitable distribution of housing among different generations using a social justice and just city approach (Corlett and Judge, 2017; Fainstein, 2010; Flyvbjerg, 2004). According to this line of argument, the Green Belt clearly *does* have problematic, negative societal consequences, reflecting Hall *et al's* (1973A, 1973B) findings in the *Containment of Urban England*. These societal conse-quences are not restricted to the Greater South East, with the Green Belt creating issues with housing needs *across* England, including in post-industrial regions such as the West Midlands (Bramley *et al*, 2017).

However, whilst the policy is widely recognised as *a* cause of the housing crisis and an element of reform is arguably needed, the evidence is not incontrovert-ible on this social justice question, and the complex reality of the housing cri-sis emerged, meaning that the Green Belt should not be abolished. For example, regarding the same question about the housing crisis, this young planner from a national housebuilder replied:

> **That is a very difficult question partly...**There are obviously are a number of other factors, but the Green Belt particularly, if you look at London and the South East (areas like Tandridge) – incredibly high house prices building very few homes and certainly the areas that we would expect to sell slightly

higher house prices happen to coincide with areas that have very limited growth. However, I don't think it is fair to say that the Green Belt is the only reason for that…[demand] is contributory…I don't think you can discount the general lack of availability of houses…some houses built in the cities haven't necessarily improved affordability.

Similar, nuanced views were expressed by other planners, who highlighted national factors, such as the economic cycle and lack of social housing, and local ones, like locational characteristics, affecting housing:

It is an outcome rather than a driver…**I don't think that you can say because of Green Belt, house prices go up – it is because we don't build enough houses. People still want to live there because they are in accessible and sustainable locations**…If you didn't have any of these other policies, the most expensive houses would be those that are in the best places where most people want to live [in the Green Belt]…with the best environment and all sorts of things. **And if you didn't have Green Belt, would they still be popular? Yes!**

(Planner, HBF)

I find it difficult to point at any conclusive evidence on the link between Green Belt as a planning policy and, let's say, residential house prices. I don't know whether the economics are that simple…there are so many factors at play…**where jobs are or what the environment is, whether there is any difference in terms of accessibility or locational factors.**

(Private Sector Planning Director (South East) (1))

Private sector planners were expected to be highly critical of the Green Belt, as assumed by neo-Marxian literature (Short *et al*, 1987), but, whilst clearly there was criticism, planners thought more broadly and critically about the housing crisis. The HBF planner's quote and others highlight that the Green Belt often covers very attractive places to live with excellent 'locational characteristics', i.e. good schools, communications and environment (see Pendlebury, 2015). Establishing to what extent these specific locational characteristics were responsible for high house prices as opposed to the Green Belt designation is very complicated as they are probably deeply intertwined (Hilber and Vermeulen, 2014).

This featured strongly in the West Midlands, where the Green Belt covers wealthy areas, such as the Meriden Gap, Bromsgrove and Lichfield, whilst the conurbation is largely 'industrial' (Goode, 2022B). As part of the research, it was found that the house price affordability ratio of a range of 12 settlements located in the WMGB was 10.9 and outside it was 7.8[2]. Regional respondents were asked whether this was evidence that the WMGB led to higher house prices as opposed to wider factors, such as locational characteristics. Most planners argued it was a mixture:

There is in Green Belts an elitism in so far as they are currently located next to the most prosperous areas. I think a lot of people who are defending Green Belt are doing it because they want to protect their own way of life. If you're in a suburban area and have immediate access to the countryside, or living in the countryside itself, you are [protecting your] own property values and the uniqueness of having Green Belt around you. I think that applies more in London and the South East, where there's a much wider Green Belt (Kent, Hertfordshire and so on). **But I think it also applies in the West Midlands. So, in large areas around Birmingham and the Black Country, are swathes of Green Belt land areas which extend as far as Warwick and areas to the east of Meriden Gap (in particular) – very sensitive enclaves for very wealthy people**. Having said that, they have to live somewhere.

(National Private Sector Planner (2))

There were similar responses from other regions, including that many places in the Green Belt are already desirable to live but the policy adds an extra 'layer' to house prices through the perceived protection it gives against development, new neighbours and views being interrupted.

Interestingly, campaigners, rather than completely dismissing the Green Belt's impact on house prices, acknowledged there was *some* impact but highlighted complexity, property stock and locational characteristics, especially in the West Midlands:

It is such a difficult question...but I don't think it is a straightforward, simple causal relationship between Green Belt policy and house prices. I mean let's take an example of Solihull...house prices in Solihull are significantly above the regional average for a particular type of house. Obviously, first of all, you have got to look at the mix of housing in each local authority area and not simply say – 'Well, house prices in Solihull are simply higher than house prices in Cannock'. Probably the main reason why that is the case is that Solihull has a lot more 4-bedroomed detached houses than Cannock does in proportion to its housing stock. **But, once you have standardised for that factor, I think purely the Green Belt has some effect on house prices but I don't think it is the only factor – it is part of the overall environmental offer.** The fact that Solihull is as it is, is partly due to it being protected by Green Belt for 60 years but also other environmental aspects.

(Planner (1), CPRE West Midlands)

Unsurprisingly, these points were also highlighted by a local politician who highlighted locational characteristics:

I think you could build all over our Green Belt and you would probably find people still want to come and live in Solihull... Obviously, the Green Belt is a contributory factor but many choose to live in the urban areas of Solihull, not necessarily in the Green Belt countryside so, you need to distinguish the

Green Belt from a control mechanism, and Green Belt in terms of its effect on the price of land.

(Local councillor, Solihull (individual views))

These quotes elucidate the complexity of the housing crisis generally and the impact of the Green Belt upon house prices, specifically with prices determined by *wider factors* than just the Green Belt, including structural forces and locational characteristics, especially in the West Midlands. This has significant implications for the broader economic literature, which often highlights planning systems as a key cause of housing crises in many countries yet struggles to disaggregate the specific impacts of policy on house prices (e.g. Ball *et al*, 2014; Hilber and Vermeulen, 2014). Nevertheless, a consensus emerged that the Green Belt is a *significant* causal factor in high house prices *in particular areas*, especially in the Greater South East and around large conurbations, such as Birmingham. Consequently, exploring the possibilities of Green Belt reform *is* important, although the extent of its responsibility alongside whether reform would 'solve' the housing crisis is also fiercely debated.

These findings have therefore underlined spatial variation as vital in understanding housing crises with markets behaving differently across countries and regions as characterised by spatial 'nuance' whilst the 'porosity' of 'local' markets to international and national investment have all been highlighted by scholars (Gallent, 2019; McGuinness *et al*, 2018; McKee *et al*, 2017, p. 60). This underscores the importance of multi-scalar factors from locational characteristics to global structural economic forces (McGuinness *et al*, 2018; Payne, 2020). The next section briefly explores other causes before critically evaluating the potential success of Green Belt reform.

The Housing Crisis: A Complex, Multifaceted Crisis – Critically Considering Other Causes

The Financialisation of Housing

Wetzstein (2017, p. 3159, 3174) wrote of the deep 'megatrends' affecting housing, including the 're-urbanisation of capital and people', 'cheap credit' and 'intra-society inequality' and described the *global* 'urban affordability crisis' so housing is affected by international, structural forces rather than *just* the *national* Green Belt. This can be related to neoliberalism, global flows of capital investment into property and its financialisation as an asset due to the rapid expansion of mortgage finance since the 1980s (Inch and Shepherd, 2019). Participants referred to these trends, including the national Conservative politician, although, given the dependence of the economy upon mortgage debt and rising house prices (Gallent, 2019; Rae, 2016), it is difficult to see how, in the short term, the economy can move away from this as Planning Academic (2) argued.

Societal and Demographic Changes

There are the large-scale, societal changes of recent decades in many countries, with a retired, long-standing West Midlands MP reflecting on how changes in household structure have increased demand for housing. Planning has to largely respond rather than shape these long-standing societal trends in a capitalist, liberal democracy, but they clearly have an important impact on housing demand (Breheny, 1997).

Existing Stock and Brownfield Land: The Urban Compaction Debate Revisited

CAMPAIGNERS PERSPECTIVES

Concerns were raised about better utilising the existing housing stock and brownfield land. This view was put most strongly by the regional policy expert (West Midlands):

> **The Green Belt is the answer to a different question…irrelevant to the housing crisis…completely irrelevant. Just a separate question, separate set of issues, separate answer**. In terms of the housing crisis as it affects the people who are hit by it, it is about social housing stock, the quality at the poorer end of the existing stock and private renting rules, regulations, and supervision. And it is the whole business about creating an urban fabric and a place-making thing…all those things hang together. It is schools, hospitals and public transport systems. The Germans and French do this better than us.

Although this was an extreme view, the importance to the housing crisis of existing housing stock was widely acknowledged as a vital factor, as strongly expressed by SSGB campaigners. For example:

> **They say there is a housing crisis but figures suggest there are as many as 250,000 houses lying empty in the country. Surely a better strategy would be to get those houses moving and make them habitable?** Couple those houses with brownfield sites and Green Belt will not have to be touched at all.

> Green Belt housing is nearly always in areas where top of the market property is built. That's not what is needed!

These quotes are particularly strong because Birmingham and the Black Country are popularly perceived to have large areas of brownfield land (Barber and Hall, 2008; Law, 2000). However, this argument was also made by some retired planners, including Strategic Planner (1), West Midlands, who juxtaposed the situation in the UK with the USA.

> **Retain a healthy market within the inner city… so that people also invest in the existing housing stock**. The great danger is that you can end up with unplanned decentralisation. So, you get this movement out and under-investment in the inner part, which will eventually lead to, if you are not careful, **the American scenario obviously which is like a hollowing out**.

However, as alluded to in the SSGB quotes and explored in Chapters 3–5, whilst *theoretically* there may be enough brownfield land and empty homes/bedrooms to meet housing needs in England, the *reality* in a capitalist economy is that people 'vote with their feet' in choosing where they live (Breheny, 1997, p. 216; Dorling, 2015). The counter-argument, espoused by Labour's 'Urban Renaissance', is that if housing stock and the quality of place in deprived areas with brownfield land is improved, such as the Black Country, people will move to those areas (Tallon, 2013). 50% of dwellings constructed nationally were flats in 2007, whilst in Birmingham, 78% of dwellings built were apartments in 2018 (Best, 2019, p. 4; Cheshire, 2014, p. 1, 2024). This demonstrates the enduring significance of planners as 'market shapers' through Urban Growth Boundaries (Adams and Watkins, 2014, p. 44) and these apartments generally cater well for young professionals and the elderly. Nevertheless, most people in many Western countries, especially families with children, prefer having a house with a garden and living in quieter, semi-rural suburbs, which was reinforced by people revaluating their housing situation and desiring more domestic and outdoor space during Coronavirus and the lockdowns (Breheny, 1997, p. 214; Goode, 2022B; Nathan, 2007, p. 5).

In turn, this raises several issues as there is currently limited public money available for brownfield regeneration and improvement of the existing property stock, whilst private property rights probably limit the scope of possible state intervention as, in a democratic, capitalist society, the State largely cannot tell or force people where to live (Breheny, 1997). Moreover, there is a range of ethical issues with estate regeneration and 'clearing' old properties, such as New Labour's Renewal Programme (Tallon, 2013, p. 91).

PLANNERS PERSPECTIVES

The viability of brownfield sites and the market was highlighted by private sector planners:

> During the nineties and early noughties, a lot of brownfield was regenerated. So, all the stuff that is readily doable, for the most part, has been done... brownfield isn't always situated in the most sustainable location. If you look in Birmingham, there is an awful lot in Digbeth that has been done and the fringes but that is very high density or flats essentially. **And, whilst that suits certain people and a certain demographic, it is not going to suit everybody...And I think it is naïve to say that urban living is the solution and over-densification**
>
> (Planner, housebuilder (2))

These arguments surrounding urban densification have become increasingly poignant since the Coronavirus and the lockdowns (Goode, 2023). One planner explored the argument, increasingly used at Green Belt/greenfield appeals in industrial cities such as Leeds and Birmingham (Young, 2020, 2023), that developments on brownfield land often yield less affordable housing due to viability:

The market goes by what is available. So, if we have got a restrictive Green Belt, does that push development to the less desirable areas?…if it is a brownfield site that perhaps is contaminated or there are other constraints that affect viability, then the developer may not be able to afford the S106 contributions towards the form of housing that is needed, education and that kind of thing. Whereas maybe, if we were on a cleaner site, they could afford to do those things…So, it is a bit of a balance on that end really.

(Private Sector Young Planner (West Midlands) (2))

Based on the evidence, it is vital that greater importance is given to the existing stock and that a brownfield first policy is retained. Nevertheless, it is apparent that the Green Belt *does* need to be 'looked at' and, probably, more land is needed for housing in the Green Belt to solve the housing crisis. Critical consideration is therefore needed as to *how* it could be reviewed, as Chapter 9 charts, with Planner (3), CPRE West Midlands (individual views) acknowledging:

I think there isn't quite enough [brownfield] land [to meet Birmingham's housing needs]. It is important to look at our existing housing stock and I do think we need to build council houses. I don't think you should just keep chipping away at the edge of the Green Belt…you should jump it and do some new towns…Redditch was a great success, in my opinion…it has worked and in the right location that could work again.

Social/Affordable Housing and the Land Market: A Key Battleground

Affordability was a key issue, especially the importance of social housing, around which a consensus formed among planners and campaigners. For example:

It would be over-reacting to blame the Green Belt entirely for the housing crisis because there are so many other hugely systematic issues in terms of our land market. **To be honest, [it] probably plays a relatively small role as opposed to all kinds of other things that might have affected affordability and the housing market, whether it is the [lack of] social housing or…the way that land in valued**…they are such big drivers and it is [Green Belt] probably blamed too much because it is sort of easy for these other people, who don't necessarily understand the complexity of housing markets which is very perplexing.

(Young planner, environment consultancy)

Again, this shows that, whilst Slade *et al* (2019) have argued that younger planners are generally given insufficient space for critical reflection, some planners thought critically and reflectively about the housing crisis. Nevertheless, whilst social housing commanded widespread support, Retired Strategic Planner (3) (West Midlands) cautioned about the difficulties of implementing it, reflecting the theme of campaigners desiring to protect an area's semi-rural 'character' (see Chapter 7):

People start to think, 'Oh, I don't want all these poor people who misbehave and it will all bring drugs and things like that and criminal activity – blah, blah, blah'. **So, a lot of it comes down to a 'them' and 'us'…they just feel it is an invasion even of their space/lifestyle.**

Moreover, there is still the crucial issue of *where* social housing would go with research on council housebuilding identifying land as a key issue (Morphet and Clifford, 2019) and the literature highlighting that affordable housing is still controversial such as Davison *et al*'s work (2016, 2017) in Australian cities. Consequently, whilst *more* land is probably needed for housebuilding, affordability is a key issue with growing recognition of the inability of volume housebuilders to cater for *all* housing needs utilising the study's definition of affordability for 'success' in solving the housing crisis (Hudson and Green, 2017). This involves the wider issues of the land market, the cost of new houses built and the oligopolistic structure of the housebuilding industry (see Chapter 4; Ryan-Collins, 2021). Again, there are aspects of these problems in other countries, especially the dominance of volume housebuilders (O'Callaghan and McGuirk, 2021), but they are particularly marked in the UK (Adams and Leishman, 2008; Barlow and King, 1992). The cost of new housing built in the Green Belt was highlighted as a major issue by politicians and campaigners, especially the attractiveness of development sites in the WMGB compared to 'brownfield/industrial' ones in the conurbation, as the SSGB campaigner quotes below demonstrate with 'affordability/affordable' being one of the most used words (14 times):

Affordability does not seem to change when many of these houses being built are at the higher end. Building £500k plus houses does not convey "housing crisis" to the vast majority of people.

Affordable housing, I see, is commonly used regarding development proposals in green belt and brownfield sites. Is this smoke and mirrors/lip service to appease…? The percentage of people who can actually afford these homes is very low. Unfortunately, when development is allowed on Green Belt land it is often for 'Executive' homes which command a price premium that takes them out of the financial reach of those who need affordable housing.

Similar observations were offered by a Project Fields campaigner who highlighted that *'new houses built here sell for over £300,000'* and a Former West Midlands Conservative MP:

Green Belt land can be very expensive so it is hard to deliver affordable housing. The land price is high because Green Belt is usually very attractive greenspace and it's at a premium…some of the most expensive land…so I think you continue to need Green Belt protection but we probably need more resources to make it possible to regenerate the brownfield sites.

Although it is easy to criticise these campaigners as vilifying volume housebuilders, 'NIMBYs' and using superficial, 'legitimate' reasons to obscure their 'true' motive of opposing development (Rydin, 1985; Short *et al*, 1987), there does appear to be genuine popular concern centred on affordability, especially that new housing in the Green Belt is too expensive 'for them', i.e. for locals (Bradley, 2018; Gallent, 2019). This was reflected in some of the quotes by retired planners. For example:

> If you took that [Green Belt] protection away, the market would actually carpet [Solihull] and all the clever developers and landowners would pile in to try and make their money out of their land and, actually, there is only a finite infrastructure environmental capacity in this area
>
> (Retired Strategic Planner (2), West Midlands)

This is reflected in a recent article in *The Economist* (2024) arguing that, even if the Labour Government initiates a significant housebuilding programme, this probably would not significantly reduce house prices.

Interestingly, planners from housebuilders and the private sector acknowledged that their aim was *not* to lower house prices, locally and nationally, because their business model is premised on maximising profit, the price of new housing is largely set in the 'second hand market' and land is expensive (McGuinness *et al*, 2018; Saunders, 2016, p. 42). Indeed, in one sense, regulation is often vital for underpinning stability and certainty in a (development) sector which is characterised by risk and volatility (Raco *et al*, 2018, 2019). Firstly, the policy can benefit housebuilders as, through restricting new housing supply and maintaining house prices, it prevents the damaging over-supply of new housing and crises of over-accumulation, which is often associated with housing market crashes in countries with weak planning systems, such as Spain and Ireland during the Financial Crisis (O'Callaghan and McGuirk, 2021). Secondly, the Green Belt and planning regulations potentially exclude smaller housebuilders both directly due to the cost and complexity of regulation and indirectly through raising the cost of land.

This can create an insurmountable barrier to entry, with the broader international literature demonstrating the impact of planning regulation on the size of housebuilders and planning consultancies (Ball *et al*, 2014). However, they highlighted that new housebuilding is still vital to provide a *proportion* of affordable housing in the absence of extensive social housebuilding, to enable 'movement' in the housing market (i.e. increasing options for first-time buyers or allowing older people to downsize) and to diversify housing stock (as explored in Chapter 4, especially Bramley's *et al* (2017) research). For example:

> **OK, so when you are buying land in the Green Belt, you always pay a premium because generally the market area is already quite high because of the lack of stock and people are desperate to buy there when they have got the money. So, your second-hand market normally dictates...** When you then build brand new houses, the expectation of the landowner is that 'Well, the market is really good', so their expectation of what the developer will pay is significantly increased and...once you have got a

consent in a Green Belt authority, you then take your units to market and there is not as much competition. So, people are willing to pay more…there are a lot of landowners who would love to sell you land in the Green Belt but their expectation of value is high and that is just because of the Green Belt impacts.

(Planner, national housebuilder (2))

This section has established how volume housebuilders *alone* cannot deliver sufficient quantities of affordable housing to solve the housing crisis, as international experience demonstrates (Bradley, 2023; Griffith and Jefferys, 2013; Morphet and Clifford, 2019). The issue seems compounded by the land price and structure of the housebuilding industry, demonstrating the value of other providers, especially social housing, to fill the 'gap' in housing provision whilst affordability and the land market need to be addressed *alongside* any reform to the Green Belt. Indeed, this proposal has gained currency in policy circles, with one of Labour's 'golden rules' for development in the 'grey belt' being a target of 50% affordable housing although this has been reduced to 15% more affordable housing than the local planning policy (Martin, 2024, p. 2).

As land prices are very important in the cost of housing, especially with the perceived scarcity of land with planning permission for residential development, the complicated and contentious underlying issue of land value capture needs addressing *alongside* reforming the Green Belt. Better ways of funding local services and providing affordable housing need to be found, and experience from the Netherlands and elsewhere demonstrates that there are more effective mechanisms of land value capture, although this is beyond this book's scope and requires wider societal and professional debate (Cheshire, 2009; Crook and Whitehead, 2019). The idea of Development Corporations to capture land value uplift in the Green Belt has recently been articulated (Cheshire and Buyuklieva, 2019; Cheshire, 2024), whilst Wolf (2015A, p. 2) suggested a 'Green Belt tax' to support brownfield development and Balen (2006) that such a tax could be used to fund widespread forest planting. A planner in Focus Group I called for land to be 'nationalised'. Although this is very contentious, there is potential for more effective land value capture accompanying Green Belt development to potentially support brownfield regeneration and improve the remaining Green Belt.

Broader Reflections on Solutions to the Housing Crisis

Reflecting upon the data, there is arguably no single 'magic bullet' to solving the housing crisis and, although Green Belt reform needs to be an important aspect of it, there are other, broader solutions, which can be learnt from international experience which space does not permit to fully explore here. This includes greater state involvement in site assembly (Letwin, 2018), as in the Netherlands, dividing up large sites between housebuilders, such as Fairfield Park, Bedfordshire, and encouraging a wider mix of housebuilders and tenures through the local plan process (Adams and Leishman, 2008; Lichfields, 2016). In terms of addressing solutions to affordability, a radical suggestion by planning academic (2) was to reduce property speculation by limiting '*all housing built... to single family occupancy*'. Whilst this idea may be theoretically desirable, there would clearly be difficulties implementing this in a capitalist economy with its viability concerns

Nonetheless, planning could have greater power to determine the tenure and type of property during the permission process. The possibility of land value capture has been raised and could be in the form of a Green Belt 'tax' which funds brownfield regeneration to address popular fears about development in the Green Belt being at the expense of brownfield land (Wolf, 2015B, p. 2). Integrating public transport with housing is key to achieve carbon neutrality alongside urban design and placemaking to ensure that developments generally, but particularly in the Green Belt, have a sense of place (Harris, 2020). Many quotes could be given but Retired Strategic Planner (3), West Midlands, was especially helpful:

> **If there are Green Belt incursions to be made, it shouldn't be on the basic of what I call 'the bog-standard developer approach'...It has to be done in a very innovative way so that it is seen as adding real value to what the place should be, look like and feel. That is why I mentioned the Hammarby Sjostad...**There must be some clear principles established as to what it should look like...ideally, we should be basically mandating them [housebuilders] to build to at least level 4, potentially 5 [in New Labour's Code for Sustainable Homes],...those are the big challenges and I think it is something that, you know, a region could take some political leadership on.

This first half of the chapter has evaluated the extent of the Green Belt's responsibility for the housing crisis and analysed who gains and losses from it using a social justice approach. Whilst the Green Belt is to some extent responsible for the housing crisis and there are 'winners' from it (homeowners through adding *some* value onto property prices), the complexity and multifaceted nature of the housing crisis means that it cannot be said to be the *primary* cause of the crisis. Conversely and paradoxically for neo-Marxian analysis, campaigners against Green Belt development are often concerned about *affordability and the expense* of new housebuilding. As Chapter 8 develops, it cannot be said to be the *sole cause of* unequal power relations in planning or unproblematically subject to 'rent-seeking' by homeowners (Rydin and Pennington, 2000, p. 153). This informs the next section evaluating the extent to which reforming the Green Belt would help reach the aspirational objective of affordable housing for all.

Does/To What Extent Does the Green Belt Need to Be Reformed to Solve the Housing Crisis?

Conceptualising Green Belt Policy Reform and Wider Recommendations

Three broad schools of opinion emerged regarding reform, although these inevitably overlapped (Figure 6.1). Most planners and campaigners agreed that the

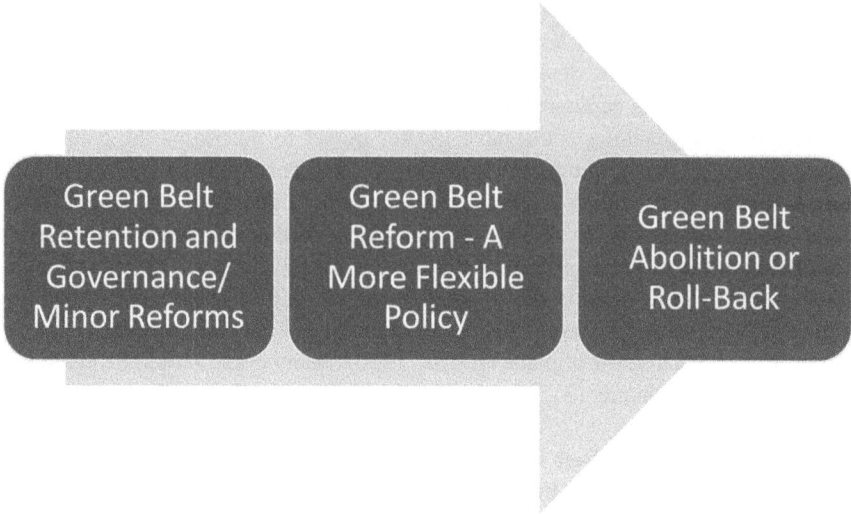

Figure 6.1 The spectrum of Green Belt perspectives.

Green Belt's governance needs reform and movement to a higher strategic level (see Chapter 9). Some planners, but mainly campaigners, argued that the *policy itself* fundamentally was working well but that there could be improvements, such as improving recreational access. Most planners, especially private sector ones, argued that the *policy should be retained but reformed and become more 'flexible'*. This would mean critically considering its overall purpose *alongside* reviewing its spatial extent (i.e. assessing all land in the Green Belt against the five purposes). Thirdly, a small minority of planners argued for its abolition or significant roll-back. Each school of thought is evaluated in the light of international experience, but Green Belt abolition is ruled out, not only for planning reasons but because it is politically unrealistic.

The strength of feeling about the Green Belt means that reform is challenging to write about (Bradley, 2019A; Sturzaker and Mell, 2016). However, consensus crystallised among planners on the need of *retaining the Green Belt's concept but reviewing its overall spatial extent and purpose(s)*. The policy improvements suggested by campaigners are adopted in this study, although the evidence suggests that it is probably time for reviewing the Green Belt, ideally as part of a national plan, as more land is needed for housing. Its purpose and spatial extent should also be debated through a national conversation (see Chapter 7; Broadway Malyan, 2015; Parham and Boyfield, 2016), with this section outlining an alternative purpose anchored around dual sustainability and social objectives.

As a national policy, its spatial extent also probably needs reviewing in broad terms nationally alongside a strategic review, such as the 1964 South East plan (Lainton, 2014), to assess how well land serves its purpose(s) and whether alterations/additions/deletions are needed. However, any reform to the Green Belt *policy* should be *alongside and in tandem with reforms to its governance structure* (see Chapter 9).

Green Belt Retention and Governance/Minor Reforms

The argument for retaining the Green Belt was made on several levels from the amount of political and popular support it commands through its historic longevity and effectiveness. Those supporting retention did not identify it as a *principal* cause of the housing crisis, with these responses elucidating these arguments:

> If you did away with the Green Belt completely, I think you would take away one part of the planning system that most people know and one of the most important parts…it wouldn't be long before the developers were saying you can take away this bit and that bit and the entire planning system would collapse. It is that deeply engrained and, also, politically almost impossible to get rid of in its entirety…I said that the policy was sound and it is, but in a sense, it is negative and I would like to see it expanded to be positive and being considered as a major regional contribution to combatting climate change…to see emphasis given to the protection and creation of more natural habitats – the more, better, bigger and joined up philosophy that Professor Newton put out.
>
> (Retired Strategic Planner (4))

Many of those arguing for 'retention' were participants from the West Midlands, where the WMGB has successfully contained a relatively compact, contiguous conurbation, and the WMGB's predominately rural character forms a juxtaposition to the 'industrial' conurbation, so it is generally well-used for recreational, such as the Lickey Hills (Goode, 2022B; Law, 2000)[3]. The quotes above are helpful in articulating a positive vision for how the Green Belt could be improved, illuminating a potentially renewed purpose (see also Kirby and Scott's work on a 'multifunctional' Green Belt (Kirby and Scott, 2023; Kirby *et al*, 2023A, 2023B, 2024). As outlined in Chapters (3 and 4), Howard and Unwin envisaged the Green Belt as an active, dynamic area for the recreational use of cities, forming the rationale for the inter-war land purchases, although this link with recreation was weakened in the post-war era with modernist ideas of separating rural and urban (Amati and Yokohari, 2003, 2004, 2006). These arguments have particular relevance and currency in the contemporary context with the recent growth in research on green infrastructure and urban nature as crucial to both climate change mitigation and urban well-being, whilst the political ecology literature has explored the capitalisation and exploitation of nature (Malik, 2024; Mell, 2024; Pykett, 2022; Reyes-Riveros *et al*, 2021; Walker, 2007). A strong case exists today for again giving the Green Belt a clearer, recreational role in its purpose(s), but private landownership alongside agricultural requirements poses challenges (although there may be a possibility of changing land use in the context of the post-Brexit subsidy regime (Kirby and Scott, 2023; Kirby *et al*, 2023A, 2023B, 2024; Harrison, 1981; HM Government, 2018)). The lockdowns accompanying Coronavirus demonstrated the pressing need for *local* greenspace and not relying too heavily on the Green Belt for the recreational use of cities (Goode, 2023; Mell, 2024).

The Green Belt needing to have an environmental purpose emerges strongly, especially the importance of natural habitats and their ecological value as reflected in a large body of literature on the 'ecosystem services' of green infrastructure (For example, Kirby and Scott, 2023; Kirby *et al*, 2023A, 2023B, 2024; Mell, 2024; Mensah *et al*, 2016, p. 150; Roberts, 2017, p. 13). An overarching objective of sustainability for the Green Belt would work well to capture these positive benefits as opposed to the negative role of preventing urban sprawl. These exciting and innovative ideas demonstrate the need for a broader, societal debate on the Green Belt's purpose, and the quotes also make the case strongly for retaining it *as a concept*. Interestingly, even the National Private Sector Planning Director (3), who argued for the Green Belt to be abolished, conceded that they would be more valuable if they had greater recreational access.

Green Belt Reform – A More Flexible Policy

Planners critical of the Green Belt sometimes lacked clarity as to how it could be reformed. The consensus seemed that, whilst the Green Belt has been successful at achieving its historic objective of preventing urban sprawl, they were critical of whether these objectives were 'right' and appropriate for the twenty-first century, stressing the need for a more 'flexible' Green Belt with this phrase, 'flexible/flexibility', being constantly used. Raco *et al* (2018) similarly found the importance of 'flexibility' to developers in city centre/inner city developments, with this reflecting broader critiques of planning systems being outdated as originating in and reflecting post-war Fordist economies (Prior and Raemaekers, 2007). Secondly, concerns were raised about if the Green Belt's current spatial extent is 'right' and whether all land serves the Green Belt's purposes.

The point was made that *a* Green Belt was and is necessary, especially around London, and that the extent of Abercrombie's Green Belt was generally 'right'. The large-scale, further extensions to the Metropolitan Green Belt in the 1970s/80s meant that it was now too 'large' although planners largely did not comment on the size of what the WMGB should be, thus reflecting its smaller spatial extent. Thirdly, it was often highlighted that the Green Belt was introduced alongside new towns and regional policy, so their effective working depends upon other mechanisms for meeting housing needs, whether new towns or urban regeneration. However, it was argued that the Green Belt has now become an entity in itself rather than viewed as a suite of policies, whilst their inflexibility partly stems from their political sensitivity. For example:

> **It would be helpful for national government to identify broad locations for growth. That would take a more flexible approach to the Green Belt and perhaps encourage other areas to release more Green Belt...Originally, it was to restrict urban sprawl and it particularly focussed on**

London but that does not necessarily mean that growth is now not appropriate in other areas. I feel that the Green Belt has fulfilled its purpose up to a point but there should be probably a greater review as to why certain areas have continued to be.

(Young Planner, National Housebuilder)

We need to rethink what the Green Belt is…for the 21st century, and I think that purpose should be around green infrastructure and natural capital, which is a massive agenda issue at the moment for the most. So, we need to rethink about the greenspaces we have between cities and between towns…as well as sort of the purpose of the Green Belt

(Strategic Planner, South East)

Building on these observations, modifying the Green Belt's purpose to bring sustainability to its centre, underpinned by the twin pillars of social and environmental sustainability, would be an appropriate way forward to assess whether the land is suitable for the Green Belt or affordable housing developments. However, this must go alongside a national debate/conversation on its purpose and spatial extent (see Chapter 7).

Green Belt Abolition or Roll-Back

Building on the neo-Marxian, economic framework (Short *et al*, 1987), private sector planners, especially those acting for housebuilders and landowners, were expected to be extremely critical of the policy and desire its abolition. However, there was a surprising consensus *against* its abolition and in favour of retaining the concept. This was partly related to its popular, political support so that, even if planners *personally* preferred its abolition, they did not consider this a viable option. For example:

Green Belt is so well-recognised, if it were done away with it would be seen by a lot of people as being a deliberate ploy to concrete over the countryside. So that to the extent that Green Belt has got to survive in some form, a reformed Green Belt policy that does the job it was originally intended to and is flexible enough to allow for stable plans of growth, could be achievable and deliverable. **But in my personal view, not the company view, I'd do away with it tomorrow.**

(Private Sector Planning Director (West Midlands) (2))

Private sector actors being in favour of regulation, like the Green Belt, echoes Raco *et al's* (2018, 2019) work on large-scale developments in inner London and suggests that housebuilders/planners and the private sector take a more nuanced view towards regulation than that presumed by the neo-Marxian framework. Moreover, arguably the Green Belt benefits housebuilders to some extent through restricting new housing supply, maintaining house prices and excluding smaller housebuilders due to the cost and complexity of regulation (Gilmore, 2014, p. 2, 2016, p. 6). Likewise, the difficulty of getting land released from the Green Belt to some extent protects their professional expertise and creates a 'need' for planners (Airey and Doughty, 2020). National Private Sector Planner (2) argued:

> It's not in anyone's interest to dispense with Green Belt altogether. I think Green Belt is one of the important parts of British planning and **it wouldn't be in the builders' interest to dispense with Green Belt as, once you do, you lose the quality of the environment. By releasing so much land altogether, you devalue what you are trying to achieve so no builder or landowner would want all the land released because otherwise that reduces its value.**

This resonates with Dear and Scott's (1981, p. 13) observation that the Green Belt is a 'historically-specific and socially-necessary response to the self-disorganising tendencies of privatised capitalist social and property relations...in space'. Indeed, regulation arguably plays an important role in many aspects of economies but is particularly vital to the property industry with planning systems in many countries managing how land is developed and the lack of robust planning regulation being highlighted as a key factor in the housing market crashes in Ireland and Spain during the Financial Crisis (Hamnett, 2009; Prior and Raemaekers, 2007). Indeed, Adams and Watkins (2014, p. 1) have demonstrated the key role that planners and the planning system play in shaping, regulating, and stimulating property markets. This suggests that a more complicated relationship exists between business, especially volume housebuilders, and planning regulation than the existing planning/geographical literature suggests (Raco *et al*, 2018, 2019).

Nevertheless, a minority of planners did advocate the Green Belt's abolition. These tended to be younger or middle-aged, and whilst some planners were quite conservative and did not think outside their professional mindset or the current legislative framework, others were more free-thinking and radical[4]:

> So many authorities around the UK don't have Green Belt but still manage to restrict development on the edges through the use of other policies...it is a bit outdated and, actually, there are other ways that you can plan proactively to restrict growth around your settlements...through a proactive plan process... maybe you don't need Green Belt at all. **I know it wouldn't be popular and I don't think a government would ever want to suggest it.**
>
> (Planner in Focus Group II)

> The purposes...of the policy [could move] toward something along the lines of more general criteria, so perhaps Biodiversity Net Gain. So, it has effectively the same principle...that would kind of produce the same sort of outcome but in a more flexible and probably more appropriate way. Just sort of slapping a land designation around certain areas – I don't think really achieves what we are trying to achieve in planning today. A specific set of criteria...specifically linked to the sprawl of London at a specific time...I am

not sure it is necessarily the most appropriate way forward for what we are trying to achieve with the system today.

(Local authority planner (South East))

Although it is important to think more broadly and critically about the Green Belt in this academic study, abolishing it is arguably politically unfeasible, as underlined by the planner in Focus Group II, so it is not considered as a serious option (Mace, 2018).

Regional Green Belts and Green Belt Swaps

Metropolitan and Regional Green Belts

This chapter has synthesised regional Green Belt perspectives, but differences exist *between* regional Green Belts. Firstly, the Metropolitan Green Belt is larger than the WMGB in *absolute* spatial terms, although, *in proportionate terms*, it is not significantly larger[5]. Nevertheless, the 'tightness' of *regional* Green Belts around conurbations, especially the WMGB, was often referred to as a major issue by planners and, although not as pronounced as the Greater South East, challenges still exist with regional housing crises and managing urban housing shortfalls, especially Birmingham's (see Chapter 9) (GL Hearn, 2018).

Green Belt Additions and Deletions

The possibility of adding land to the Green Belt or 'swaps' was raised by some:

Whether as part of [releasing land from the Green Belt], you have a quid quo pro to say actually, 'whilst we will be taking some land out, we will add to the Green Belt elsewhere. But in overall terms, the Green Belt will actually be getting bigger, not smaller'. But even that I think, would take an awful lot of political courage

(Private Sector Planning Director (South East) (2))

However, Green Belt additions are only feasible through a strategic Green Belt review, given that LPAs are rarely contiguous with Green Belt boundaries (see Chapter 9). Nonetheless, whilst acknowledging that the planner from the Countryside, Land and Business Association was strongly opposed to additions due to the challenges that it creates for farmers and landowners, arguably, their needs and requirements must be weighed against the *overall* welfare benefits of society of a more flexible Green Belt[6].

Green Belt TOD, Alternative Policies and Other Ways of Analysing the Policy

The debates in the quotes about the effectiveness of TOD, other policies, such as green wedges, and alternative ways of assessing the policy, like the sustainability of leapfrogging, were similar to those explored in Chapter 4. This study focused primarily on *social* justice and a strategic governance framework is arguably required to properly consider proposals like TOD (see Chapter 9).

Towards a Sustainability Purpose of the Green Belt

On balance, the evidence suggests that the Green Belt's purpose needs reviewing, but as outlined in Chapter 7, this needs to be part of a national 'conversation' and debate on the trade-offs, costs and benefits of Green Belts as part of a national plan and following a programme of public education on planning. Indeed, a retired Conservative West Midlands MP, who was very supportive of the Green Belt, acknowledged that all the land in England needs reviewing, potentially through a Royal Commission on land use. A national plan could also take a similar form to the New Town Commission or the UK2070 Commission (2020), whereby evidence is collected from a range of sources (Cullingworth *et al*, 2015).

This book recommends an overall sustainability purpose to give the Green Belt a positive purpose compared to the rather negative and perhaps outdated one of preventing urban sprawl. Focusing on sustainability would help to modernise the concept, meaning that it is more relevant in the context of rapid climate and environmental change (Kirby and Scott, 2023; Kirby *et al*, 2023A, 2023B, 2024; Monbiot *et al*, 2019). Indeed, this relates to scholarly and policy debates on the importance of green infrastructure and urban nature as crucial to both climate change mitigation and urban well-being, particularly in a post-COVID context, as well as to the political ecology literature on the exploitation and capitalisation of greenspace (Malik, 2024; Mell, 2024; Pykett, 2022; Reyes-Riveros *et al*, 2021; Walker, 2007). As Table 6.1 shows, this necessarily broad purpose would be underpinned by the interlocking pillars of social and environmental sustainability.

Environmental sustainability builds upon campaigners and planners recommendations of a 'greener' Green Belt to make it 'work harder' in its environmental value through shifting the focus from landscape/openness to more flexible environmental enhancement. This would focus on river restoration and community farming, which could be encouraged with the post-Brexit subsidy regime of 'public money for public goods' (HM Government, 2018). This would interlink with a social purpose, tied to recreation (as also supported by the RTPI (Blyth, 2017)), where these sorts of beneficial activities would be encouraged, thereby broadening the Green Belt's purpose beyond environmental to social sustainability (Sturzaker and Shucksmith, 2011).

Table 6.1 Overarching purpose of the Green Belt: sustainability

Environmental	Social
• Support biodiversity/green infrastructure. • Prevent urban sprawl. • Ecosystem services, i.e. river restoration.	• Enhance recreational access. • Support 'social' uses, i.e. urban farms/allotments • Affordable housing.

However, the affordability issue needs addressing, especially of new housing in the Green Belt, so a social purpose could stipulate that development in sustainable locations must serve a social purpose, i.e. have a reasonable level of affordable or social housing or fulfil a particular 'need' in an area, like housing for the elderly (RTPI, 2016A; Blyth, 2017). If stipulated in policy, hopefully, this would reduce landowners' expectations about the price of land and new housing in the Green Belt may become more affordable to allay the public's fears regarding affordability[7]. Indeed, this is reflected in the target of 15% more affordable housing in Labour's 'golden rules' for development in the 'grey belt' (Martin, 2024, p. 2). These points are elucidated in these quotes:

> The RTPI has said it is time to think about…a social purpose…there is a social aspect to Green Belt we can look at: who uses them? who benefits from them? There is a kind of moment with Green Belt…where you could say – 'well, if you added a social purpose, that would mean if someone were to come forward with a proposal to build 100% affordable/social housing in an accessible location of the Green Belt, politicians might say – 'Well, OK that meets social purpose''. That's a very different issue from saying that's speculative development for 100% market housing should be allowed…the Green Belt could perform a kind of unique useful function where, to a certain extent the hope value in Green Belt land is quite low (because people believe it won't be changed).
>
> (Planner from the RTPI)

> There is a growing interest in promoting woodland and wetland creation in the Green Belt on the urban fringe (in particular) and that probably has a very good mixture of public benefits. Also, improving land management for climate change and adapting to fix it…we would certainly agree that more could be done to encourage sustainable food production as well…this is what we would call the misconceptions that are put around by the anti-Green Belt people – that Green Belt has no specific environmental value because it is intensively farmed.
>
> (Planner, environmental charity)

Conclusions

Reflecting findings from other countries, the housing crisis in England is a complex, multifaceted, and long-standing problem involving a mixture of supply and demand issues, although it has intensified in recent years (Christophers, 2018, 2019; Lund, 2017, 2019). Relying largely on private sector housebuilders and/or releasing land from the Green Belt for housing *alone* will arguably not be sufficient to solve the housing crisis, especially with the dependence of England's economic model on homeownership and rising house prices (Archer and Cole, 2014; Bradley, 2023; Gallent, 2019). Indeed, countries with 'loose' planning systems, like the US and New Zealand, still experience housing crises.

However, in the absence of a fundamental change to this model, neither is the problem of affordability going to go away or solve itself. Based upon the arguments of planners and campaigners, this chapter has highlighted important proposals which could address the central issue of affordability *and* a Green Belt with a sustainability purpose fit for the twenty-first century. On affordability, a social purpose would help to ensure that more affordable housing gets built there, as reflected in Labour's 'golden rule' of 15% more affordable housing in development in the 'grey belt' (Martin, 2024, p. 1). More broadly, as reflected by experience in the Netherlands, better funding and coordination of infrastructure potentially through the land assembly by the public sector, sufficient facilities/services, and higher quality design accompanying development are vital for (re)building public trust and confidence in housebuilding (see Chapter 7). Coupled with an environmental purpose, a 'greener' Green Belt could emerge with increased recreational access whilst encouraging biodiversity and ecological improvements in line with recent research on the benefits of greenspace in terms of climate change mitigation and urban well-being (Mell, 2024; Reyes-Riveros *et al*, 2021).

However, the constant theme stressed by campaigners and planners is that the Green Belt's *purpose* should not be divorced from its *spatial extent and governance,* so it is vital that any Green Belt reviews take place strategically to properly plan any releases (see 'recommendations' Chapters 9 and 10). The data also highlighted the case for more powerful, proactive and progressive planning as a central means by which the housing crisis could be solved in the same way that the State took a leading role in development in the post-war era from 1945 until 1979 and house prices remained stable (Bradley, 2023; Gallent, 2019; Lund, 2017, 2019). The next two chapters (Chapters 7 and 8) explore the crucial issues of the social and political feasibility of Green Belt reform. As developed in Chapter 7, a national Green Belt debate, alongside more public engagement and education in planning, is vital so that Green Belt reform commands public consent through potentially involving more people in planning.

Although this chapter could have been more ambitious regarding Green Belt reform and was largely written within the parameters of the capitalist system of mass homeownership, this book's golden thread is that planning is inherently political. Consequently, it is arguably more valuable, in terms of research 'impact', to stay largely within the realms of what is realistically politically possible whilst recognising what may be theoretically desirable, such as replacing Green Belt with another policy (Martin, 2001, p. 112; Woods and Gardner, 2011; Wood, 2005A).

The chapter has also highlighted the importance of measures designed to diversify housing providers, especially building more social housing, reducing empty homes and providing more support for brownfield remediation to lessen development pressure on the Green Belt. However, it is still important to critically consider how far land in the Green Belt could be used to accommodate more housing so that the character of *existing* urban areas does not change or deteriorate too rapidly due to large-scale urban intensification and densification. To some extent, the *way or process by which* development takes place in the Green Belt is as important as *where it takes place*. The next three chapters (Chapters 7, 8 and 9) therefore turn to the *process* of planning. Finally, arguably, more research is needed to contribute to the wider theorisation of the development sector and planning regulation in different spatial contexts, including where there are weaker planning systems and popular/political cultures with less strong commitment to urban containment like Ireland and the USA. Moreover, other forms of planning regulations, such as zoning and National Parks, need more empirical examination. However, planners being generally more supportive of the principle of the Green Belt than the assumption by the neo-Marxian framework (see Chapters 2 and 5) suggest that the relationship between planning regulation and housebuilders is more complicated than the literature suggests (Raco *et al*, 2018) with the chapter highlighting the broader importance of regulation to the property industry.

Notes

1 *Tripwire* is the RTPI West Midlands Magazine and *Branchout* the South East one of England Magazine.
2 The 'Green Belt' settlements were Solihull, Hagley, Kidderminster, Wombourne, Brewood, Cannock, Shenstone, Kenilworth, Balsall Common, Bloxwich, Henley-in-Arden, Alcester. The 'unconstrained' settlements were Worcester, Evesham, Shrewsbury, Stafford, Atherstone, Daventry, Wellesbourne, Southam, Leicester, Coalville, Burton-on-Trent and Uttoxeter. The house price affordability ratio was found by dividing average house prices by average incomes (Hilber and Vermeulen, 2014).
3 The West Midlands is more densely populated, 8,380/sq mi, than the next largest conurbation, Greater Manchester, 5710/sq mi (CPRE and Natural England, 2010, p. 29). The Greater Manchester/West Riding Green Belts are also 'moth eaten' and not continuous like the WMBG (Planner (2), CPRE West Midlands).
4 The author encouraged participants to think about theoretical and ideal possibilities alongside the *status quo* in line with calls for planners to be 'reflective practitioners' (Slade *et al*, 2019, p. 8).
5 Calculations based on data from CPRE and Natural England (2010) and population statistics from the ONS.
6 A middle-way approach would be to have strategic Green Belt reviews every 25–30 years to give more certainty to landowners and farmers (see Chapter 9).
7 A Green Belt 'tax' could also dampen landowners expectations (Wolf, 2015A, p. 2).

7 Conceptualising community support for the Green Belt and opposition to housebuilding

Introduction

This chapter also proved challenging to write as the strength of feeling about the Green Belt again became evident when speaking to regional and national planning stakeholders. However, as outlined in Chapter 6, the public acceptability of Green Belt reform and wider housebuilding is vitally important to the governance and *process* of planning, especially with growing popular awareness of the housing crisis and the Government's 'target' of 1.5 million new homes over the course of the 2024–29 Parliament (Bradley, 2019A; Mace, 2018; Martin, 2024, p. 1). The chapter explores the motivations and attitudes of campaigners and communities to the Green Belt alongside broader attitudes to housebuilding as viewed by planners and campaigners/the public by weaving together qualitative and quantitative material. Although the findings are generalised nationally/internationally, they are grounded in the regional case study of the West Midlands. This is vital as housing policy in many places is largely 'place neutral' and 'spatially blind' whilst arguments surrounding the house price hypothesis are often based on broad datasets or inferences upon intuitively powerful and logical assumptions rather than directly engaging with campaigners (Bradley, 2019A, 2019B; McGuinness *et al*, 2018, p. 330).

Given debates internationally in the literature about community involvement, this chapter relates the findings to planning and geographical theory as underlining the limitations of the materialist framework of the house price hypothesis, as outlined in Chapter 4. It explores the usefulness of the place attachment, environmental psychology and cultural geography literature, which underlines popular emotional attachment to place and the countryside (Anton and Lawrence, 2014; Davison *et al*, 2016, 2017; Short *et al*, 1987). It contextualises Chapter 8 on *how* campaigners seek to exercise power in planning and outlines recommendations highlighted by planners and campaigners, thus establishing the empirical groundwork for strategic planning as the cornerstone for rebuilding public trust in planning (see Chapter 9).

DOI: 10.4324/9781032674315-7

Conceptualising Why Communities Support the Green Belt and Oppose Development: The Views of Planners

Popular Knowledge of the Green Belt

Public consultation and the public having the opportunity to be involved in decision-making has been a long-standing feature of many planning systems, which has given rise to wide-ranging debate in the literature about the representativeness and motives of those involved in participation (Bradley, 2019A; DeVerteuil, 2013). The relationship between community opposition to housebuilding and planning is complicated and the subject of extensive academic and practitioner debate (see Chapter 4) (Brownill and Inch, 2019; Sturzaker, 2017). On the one hand, a body of literature portrays campaigners as powerless as having legitimate experiential knowledge and principled concerns, such as about affordable housing, circumscribed and silenced by developers and planning systems (for example, Bradley, 2019B; Inch *et al*, 2020). On the other hand, another broad body of literature portrays campaigners as self-interested, unrepresentative, parochial and materially impacting and influencing decision-making (for example, Rydin, 1985; Sturzaker, 2010). Broader debates about the representativeness and amount of engagement in planning are largely beyond this study's scope but were reflected in discussions on the Green Belt and housebuilding generally.

However, notwithstanding the difficulties created by limitations in the public's understanding of planning, especially the Green Belt, one of the policy's greatest and most enduring achievements is its 'capacity' to interest and 'engage political public's at a time of widespread apathy with, and disinterest in planning (Bradley, 2019A, p. 181; Law, 2000, p. 57) as some practitioners recognised:

> It is probably the most positive aspect of the planning system that has entered the public psyche and for that reason it is a very strongly supported subject. Not just by organisations, like CPRE, but individual members of the public are aware of what Green Belt is. **Often, they misunderstand what Green Belt is but they cling to it with great enthusiasm. In PR terms…it is 10/10 effective.**
>
> (Retired Planner (1))

Some planners were more sympathetic to campaigners, whilst others, like Public Sector Planner (3) (West Midlands), acknowledged the barriers to people getting involved in planning, especially its technical nature and the lengthy nature of most planning documents (Parker and Street, 2015). Others were more sceptical, such as National Private Sector Planner (1), who argued that there should be greater trust in planners as 'experts', like the NHS, rather than arguing that people need training and education to engage more effectively with planning as Wargent and Parker (2018) have argued regarding Neighbourhood Planning, Monbiot *et al* (2019) in relation to 'Planning Juries' and the Raynsford Review (2018A, 2018B) about planning more generally. This shows broader, deeper levels of distrust between planners and campaigners as seen in this observation by a Policy and Campaign Advisor, West Midlands:

> People in less affluent areas are also people who know less [about] how to [articulate their] point of view but they still have the same values about the Green Belt...people believe, have their faith in Green Belt, which is good, but there is a lot of confusion between greenfield and Green Belt...It is very easy for particular politicians but, particularly for the housebuilders, to present the image that – 'because there are these people who object and care about their countryside and have a self-interest (as we all do) – all this is a tool for NIMBYism'. Therefore, it is that kind of image that is presented as, 'why should we protect these middle-class people who have got lots of money?' (They are not all middle-class who live in or on the edge of the Green Belt!)...It is an easy way of getting around the difficulties.

In planning systems, in many ways, there is a conflict or tension between 'planning' knowledge, which is associated with rationality, the expertise of planners as professionals and viewed as technical, objective or even 'scientific' and the more emotional, experiential and tacit 'community' knowledge associated with campaigners (Bradley, 2018, p. 24). Although both types are key, an important component of this book is assessing what the *right balance* should be in the planning system.

Planners Perspectives on the Motivations of Campaigners

Overview

Planners are the key actors in planning systems and have extensive interaction with the public and wide-ranging knowledge and experience of community opposition and its resulting politics. Planners' views of campaigners and their motivations may not always be representative as sometimes viewing campaigners as irrational and ignorant juxtaposed to planners being 'experts' and 'objective' (through constructing 'conceptual binaries' and resorting to the 'public deficit explanation' (Gibson, 2005, p. 383; Welsh and Wynne, 2013, p. 552)). However, how planners *perceive* community opposition is still vitally important and needs to be researched (Brownill and Inch, 2019; Inch *et al*, 2019, 2020). Community attitudes are particularly important regarding the Green Belt given the widespread public support that it commands in principle and the fierce opposition there often is to releasing land from it for housing in practice.

Planners generally viewed the public as not understanding the Green Belt policy (the public deficit explanation (Welsh and Wynne, 2013)) and were particularly sceptical and discrediting of campaigners. Mace (2018, p. 4) helpfully distinguished between 'rationalistic' and 'normative' support for the Green Belt, whereby 'rationalistic' relates to house prices and the material impacts of development for individuals or a political tool for politicians, whereas 'normative' relates to more principled, conceptual support. However, this chapter explores how most planners argued that house prices were not campaigners' *primary consideration* (the 'property' argument/house price hypothesis), but located opposition to house-building as being a deeper, more instinctive emotional fear of change. This included the specific desire to protect one's local area (the 'place' argument) and underlying popular love of, and attachment to the countryside and Green Belt as principles (the 'principle' argument). Of course, property prices could be wrapped up in desires to protect one's local area and the Green Belt being popularly perceived as synonymous with very desirable, pleasant places to live, such as Solihull/Warwick/Lichfield/Four Oaks in the WMGB (see Figure 2.1), was often mentioned.

However, opposition was often related to protecting *a semi-rural way of life rather than directly protecting house prices*. Moreover, intertwined with these normative, principled concerns, were more materialistic fears about the impacts of development on infrastructure and facilities, such as traffic and school/GP places. More research is needed on community opposition in other spatial contexts, such as gentrification, densification, rural development and affordable housing. Nevertheless, the data in this book underlines the importance of research more broadly being cognisant of space, place attachment and cultural geography in the study of campaigners as opposed to the primarily materialistic emphasis of the house price hypothesis literature (see Chapter 5) (Bailey *et al*, 2016; Davison *et al*, 2016; Devine-Wright, 2009).

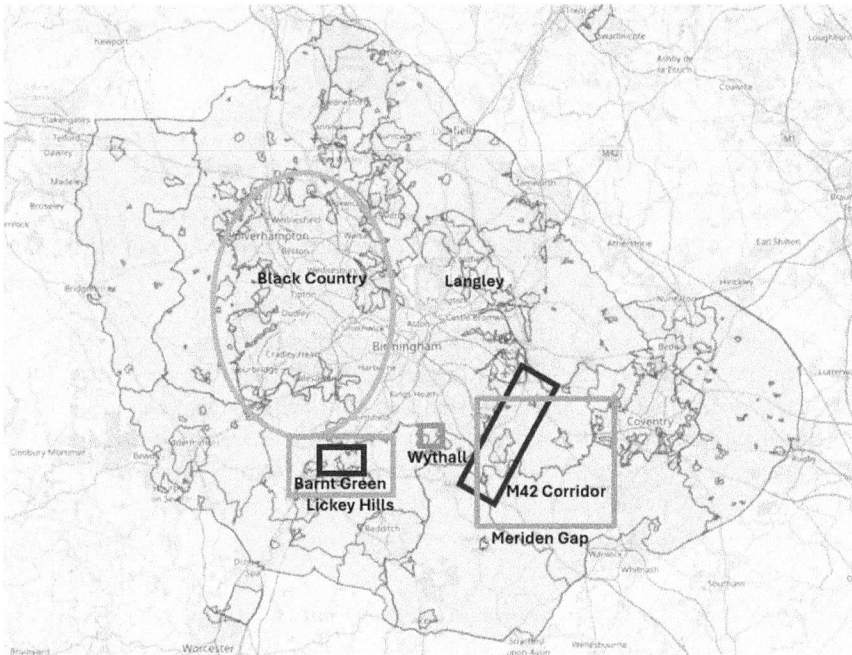

Figure 2.1 (repeated): Key locations in the West Midlands (adapted from: https://bit.ly/3rXTGuW).

Finally, it was argued that these multi-scalar fears of change among *everyday* campaigners were often deeper motivations than support for the Green Belt *per se* but, as the strongest protection against development, it was often used as an oppositional technique to serve the underlying objective of preventing change.

General Points

However, there was still heterogeneity in responses from planners whilst the difficulties of establishing general popular motivations were acknowledged, as this Senior Civil Servant in MHCLG argued:

> There have been a few surveys[1] and we should try to get to the root of this, but it is very difficult because of what people say and what is their actual reason?

Indeed, although themes emerged, planners acknowledged that people oppose development for a range of interrelated reasons and a complex web of sometimes contradictory motivations. Although this makes conceptualising and theorising more complicated, acknowledging the diversity, hybridity and intersectionality of campaigners is key as recognising them as 'real', complex people as this quote demonstrates (McDowell, 2016). This cautions against recommending one 'silver bullet' policy solution addressing and satisfying why people oppose housebuilding, but the range of proposals developed in this book, based upon the observations of planners and campaigners, seek to address these varied yet common motivations.

Motivation – Place/Principle Arguments

PRINCIPLE ARGUMENTS: SUPPORT FOR THE GREEN BELT AND THE COUNTRYSIDE

Firstly, the normative aspects of the principle/place argument were well articulated at both ends of the political spectrum. For example, a Labour Advisor on housing interviewed explained:

> The opposition from the public is essentially to development…a sort of anti-development vogue…people just don't like change and then they run a lot of arguments [like]…'Oh well, it won't be affordable to that many people', the disruption from the construction process, strain on services etc., traffic congestion, air pollution (this is an avenue they have recently taken to) – I mean, you know, they really go for it. They throw everything in… So, I think it is just change.

Support for the Countryside and Green Belt Is Interrelated

As Roy (2011, p. 591) has argued in relation to women's movements in India, alongside the 'professionalisation' of campaign groups, there is still evidence of the 'passion' related to campaigning causes and making an 'emotional' case regarding the legitimacy of their claims to knowledge. This is reflected in, firstly, planners often referring to popular support of the countryside among campaigners and the public as an intuitive response and universal principle which can be appealed to. The key to campaigning success is being able to persuade people that what is at stake is not merely a NIMBY or local issue, but a legitimate, general issue/principle that people feel strongly about (like a development being a representative 'threat' to the countryside or wider Green Belt) (see Chapter 8) (Amati, 2007; Amati and Yokohari, 2006). Planners accused the public of largely misunderstanding the difference between greenfield and Green Belt sites showing the powerful 'affective' narrative of the 'countryside' with the Green Belt being popularly seen as a synonymous institution (Mace, 2018; Warren and Clifford, 2005, p. 378). As Planner (2), CPRE West Midlands argued:

If you abolish the Green Belt, there will be just as much objection to development!

This was echoed by a young planner from a national housebuilder:

They [the public] see that countryside outside their house has just as much value as Green Belt and therefore we come up to as much objection and local opposition...to it [development generally]...**But, because the Green Belt policy has stayed similar over a really long period of time, it means that the layperson has a really clear understanding of what that means...**without much understanding of how it is possible to release land from the Green Belt under the current policy through the local plan process. [This] becomes a bit of a sticking point and [Green Belt] **is probably the most accessible, short plan policy that we have and therefore people...will stick to it very clearly**. You know, when you look at what people say at public consultation, perhaps on the local plan or large-scale planning applications, **it tends to be things that they understand. Number one is building on the Green Belt and building on the countryside which sometimes gets a bit confused...that contributes massively to how people view releases of land from the Green Belt because I don't think they understand it.**

This illustrates the central paradox that, whilst planners generally hold engagement in planning as an inherent 'good' (Brownill and Inch, 2019), there is widespread frustration with, and distrust of the public's passionate support of the Green Belt.

The Principle of Protecting the Countryside

The quotes above and other planners highlighted that campaigners' concerns relate more to the underlying desire to protect the principle of the countryside:

Whether it is Green Belt or not, we will still probably be met with the same level of objection (almost). Green Belts – people will know it's Green Belt and say 'Don't touch it, it is Green Belt, I read in the Daily Mail da di da di da'. But, obviously Green Belt can be released, there is a mechanism for it to happen and there is a development plan process and we do explain that to people…You explain to them that they don't have a right over a view, which tends to be quite an incendiary thing to say, but I don't care! They will say there are 'slugs and bugs and unicorns and fairies living on the site you know'! But honestly, we have to be super robust with all of that. But their argument, 'Well, it's just Green Belt', **even people say a site is Green Belt when it isn't Green Belt because, again, perception of Green Belt is that, if it isn't built on, it is Green Belt.**

(Private Sector Planning Director (West Midlands) (1))

The Principle of Protecting the Green Belt

Other planners underlined popular support for the policy as a *principle*, rather just an oppositional strategy although there was disagreement on whether the Green Belt or countryside was more important as a motivation. For example:

CPRE are particularly active…because **Green Belt has always been such a national narrative that everybody is sort of in with it and it is easy to pick that up as part of their representation. I think people are very canny nowadays.** Most people know how to object to a planning application – what to say, **what are reasonable planning reasons. Because Green Belt sits at the top of that list, it is always what people talk about**…People will object on transport but it is very opaque…[and] difficult for a local person to actually understand and therefore challenge effectively…The Green Belt policy is there in black and white, fairly short, it's pretty accessible and all **you need to write on your letter of objection is 'It's in the Green Belt and therefore it is inappropriate'. And that makes it a really easy mechanism for people to object to development.** [Green Belt] is where they walk their dog every day or they ride their horse or just like being in the countryside and they don't want to feel like they are going to lose out on it or that the village services are going to be overwhelmed by **lots of people who don't have this particular attachment to the Green Belt as a principle…That differentiates it from any other person that lives next to a part of the countryside.** The opposition is greater in Green Belt areas.

(Planner from housebuilder (2))

These quotes show that normative opposition is often driven by the interrelated, interlocking principles and values of the countryside *and* Green Belt, which are *both* important, inseparable factors as often conflated and confused (Mace, 2018). In a sense, it is more challenging for public policy to respond to general values and popular imagination, especially emotional support for the countryside, than direct economic motivations with this chapter recommending a national Green Belt 'conversation' (Inch *et al*, 2020; Mace, 2018).

Place Arguments: Protecting One's Local Place

These general principles intersected and crystallised with place-specific concerns, both 'materialistic' *and* 'normative', regarding people's vision of place underscoring the importance of place attachment and environmental psychology (Mace, 2018; van den Nouwelant *et al*, 2015, p. 13). Materialistic concerns pose poignant questions for public policy because, according to what planners say about campaigners, infrastructure and the effect of development on local services features prominently as a motivation:

> **The majority** [of objections to a Green Belt site] **are related to infrastructure…the most frequent ones that came up…loss of Green Belt did feature but surprisingly not as high as the very kind of tangible, real things that would have to be faced day by day**. People talk about the clogging up of roads – 'How can we take another 15,000 cars?' so transport, education, the lack of schools and health facilities. Some of them did mention ecology and the loss of trees and habitat but **the majority were related to physical infrastructure.**
>
> (Policy Planner (1) (West Midlands Council))

This was not just a concern to the West Midlands but appeared across the country as a major issue, with planners from housebuilder (2) arguing that it partly related to 'unplanned development'. Participants raised many issues with planning that are beyond the Green Belt, including the fragmentation of different layers of local Government and separation of land use planning from transport and healthcare planning (see Chapter 9) (Riddell, 2020). A young planner from a national housebuilder invoked Abercrombie and new town principles in support of large-scale developments, which she partly blamed the Green Belt for preventing.

However, these materialistic, specific concerns interwove with normative concerns and principles in the fear of change, although this was wider than just planning:

> Even if it were a non-Green Belt site, **they** [campaigners] **will still come up, trot out the usual objections because everyone is a self-appointed activist when it comes to trip generation, for instance,**[2]…they will over-exaggerate. I mean it is all utter, utter nonsense…They will object on the fact that they can't get access to their doctor at the surgery…They will suggest, and I have had this suggested to my face, that 'It's these immigrants that are causing the problem and that is why I can't get in and see my doctor'. And it is entirely baseless and quite often verging on racism as well as xenophobia. They will talk about how they can't get into the schools…It is almost an

age-old conversation that I repeat regularly, people know about the Green Belt and they will use that as an argument **but they will also trot out some standard -**.

(Private Sector Planning Director (West Midlands) (2))

In line with place attachment theory and environmental psychology (Devine-Wright, 2009; Gross, 2007), people are often afraid of their area changing in character, which is particularly poignant in the WMGB with often highly desirable, rural/semi-rural places to live, such as the Meriden Gap, juxtaposed to 'industrial' conurbations (Goode, 2022B; Harrison and Clifford, 2016). Concerns seemed not so much *direct* house prices but protecting one's residential and wider area's rural 'exclusivity', character and way of life (Short *et al*, 1987, p. 36):

> There is a clear kind of disparity created in and from the fact that there are very nice areas because they are Green Belt and nice countryside with an idyllic kind of lifestyle. It is a very British kind of thing to have that and I think that is synonymous (nice countryside) with the Green Belt, especially in those [West Midlands] villages.
>
> (Private Sector Young Planner (West Midlands)

These quotes show that, certainly in the WMGB, planners gave primacy to desires to protect and preserve one's local area as a motivation, again chiming with the place attachment and cultural geography literature (Thrift, 2004).

Uniting the Place and Principle Arguments: Fear of Change

These normative and materialistic concerns can be grouped under the broader motivation of the fear of change. Firstly, this relates to powerful binaries and the potential for one change to sequentially lead to another so relaxing the Green Belt being a pre-cursor to abolishing it which intertwines with the 'politics of affect' and generic fears about the countryside being 'concreted over' (Thrift, 2004, p. 57; Warren and Clifford, 2005, p. 361). Secondly, fear of change relates to development changing a place's character including fears of different people moving in, local infrastructure/facilities and the impact of construction work (Bailey *et al*, 2016). For example:

> **The prime thing is people don't like change. You could almost put a full stop there, because then they start looking for reasons why.** So, it doesn't always work out [house prices reducing upon development] but there is probably a fear about that…**a worry, about noise, traffic, safety and there is very definitely a fear of crime.**
>
> (Retired Strategic Planner (4))

This probably reflects the lack of control people feel that they have over their communities and the built environment, which may be reflected in the Brexit vote

(Inch and Shepherd, 2019). Whilst the Private Sector Planning Director (South East) (2) did not directly establish a 'conceptual binary' of campaigners and the public being universally resistant to change by highlighting a broader, instinctively cautious attitude (Gibson, 2005, p. 383), other planners, such as this LPA planner (South East), questioned the 'public deficit explanation' narrative (Welsh and Wynne, 2013, p. 552) of the fear of change. He argued that opposition is more about wanting to stop change, especially development, from happening whilst accusing the public of 'rash' thoughts:

> I wouldn't say fear of change… people know exactly what to expect. It is not as though it is an unknown concept out there but I am sure it is the consequences. They know what to expect and just don't like it – 'we don't want this' is the answer…The Green Belt sort of solidifies those areas of contest in some kind of specific frame of rash thought for people to attach to.

Whilst this is a valid point about people not *liking* change, and the book is focusing on a renowned planning policy in an English context (more research is needed on people's attitudes to change in different geographical contexts), this section has raised the important issue that people are often *resistant* to change and, notwithstanding different possible motives, this raises poignant questions which are addressed in the recommendations. Firstly, how effectively does public and planning policy respond to the normative concerns people have, whether at a principle or place level? Secondly, given people's material concerns, whether in the Green Belt or not, how can planning better address concerns about infrastructure and facilities?

The Property Argument

Some planners located popular concern about property prices as being the important motivation for campaigners, especially when development *directly* affects a person's property. However, it was highlighted that even developments adjoining a person's home rarely lead to *significant and long-term* falls in house prices, as Retired Strategic Planner (4) explained (see Chapter 4):

> **The house price would come in for a very small percentage of people who are more or less directly affected, for example, the end of your back garden** although perceptions of these are very difficult.

However, some planners highlighted the popular *perception* that development reduces house prices as a key motivation:

> **The majority of people are objecting on the basis of things that result in loss of value of their property.**
>
> (Former LPA Director, West Midlands)

Quite often, people have bought their house or moved to an area… because of the Green Belt and they think they are 'safe' (and they are not). **And, particularly people who are on the edge of Green Belts and have paid a premium for their view, they are probably more unsafe than anyone else…There is certainly a view that quite often it falls around – 'Well, you are taking my view away, reducing my house price**.

(Public Sector Planner (3), West Midlands)

Nonetheless, the consensus seemed that place and principle concerns were more important as motivations for campaigners.

Strategic or Localist Approaches by Campaigners?

Bradley (2019A, p. 166, 2019B, p. 695) argued that campaigners largely view the (North West) Green Belt (NWGB) as a coherent whole or entity and within the concept of the 'commons' with the history in the North of recreational access to the countryside. However, in Greater Manchester, the Combined Authority having planning powers means that there is one sub-regional Green Belt Review, so it is easier for campaigners to unite, form opposition/campaign groups and raise awareness, whereas, in other parts of the country, particularly the West Midlands, there is more fragmented planning (Bradley, 2019A, p. 694; see Chapter 9). Whilst there was strategic concern among professional campaigners, the general view among planners of *everyday* campaigners seemed that they largely took a localist perspective (deepened by the localism agenda) (Goode, 2022A):

At the moment, it is more of a Duty to undermine and confiscate than to Cooperate…Every single one of the responses on our consultation, well most of them, will be on the Green Belt, and…use the phrase, 'We don't want our Green Belt to be plundered. I don't mind the next borough's Green Belt, fine for them, it doesn't matter, but our Green Belt will always remain untouchable'. Well, it is not our Green Belt really so it is not for us – our community. **It is to stop the urban sprawl and it is the Metropolitan Green Belt and so it is a regional issue.**

(Local authority planner (South East))

Conclusions from What Planners Say About Campaigners

This section has highlighted that planners argue that most people are more concerned about principle and place factors rather than property prices, thus resonating with the international place attachment, environmental psychology and culture geography literature (DeVerteuil, 2013; Devine-Wright, 2009). This suggests, in terms of public policy, that planners and policymakers need to pay greater attention to people's emotional attachment to place, alongside material factors, such as improving facilities, rather than *purely* focusing on economic self-interest (Inch *et al*, 2020).

Conceptualising Why Communities Oppose Development: What Campaigners Say About Their Motivations for Opposing Development

Introduction

There is sometimes a gap between what people say and believe in interviews, meaning that motivations are often difficult to establish, especially as campaigners can be a self-selecting group (Lloyd, 2006, p. 10). Consequently, quantitative datasets are explored to triangulate and analyse a wider, potentially more objective sample, due to the anonymity of questionnaires (Saunders *et al*, 2016).

Quantitative Data

The datasets directly relating to the Green Belt are analysed before exploring broader datasets on people's attitudes to housebuilding.

Popular Attitudes Towards Green Belt

CPRE GREEN BELT QUESTIONNAIRE 2015

Table 7.1 shows that the policy commands widespread popular support across different regions and social groups, albeit it has less support in urban areas and among social classes (DE) and renters. However, these differences were not as significant as predicted by the house price hypothesis suggesting that the public supports the Green Belt and countryside as a matter of principle (Harrison and Clifford, 2016; Mace, 2018), rather than *just* homeowners for direct amenity or economic reasons. However, there has been a significant temporal change in attitudes because, in the 2005 CPRE questionnaire (p. 1), which asked a similar question, 84% of people supported the Green Belt compared to 64% in 2015. It remains to be seen how far it commanded *particular* support among the inter-/post-war, baby-boomer generation with CPRE (2015) arguing that the housing crisis has caused the erosion of popular support. Nonetheless, the policy continues to command widespread public support compared to other regional growth management policies in other countries, like the 'Green Heart' in the Netherlands or the Urban Growth Boundary in Portland (Altes, 2017; Layzer, 2012).

However, despite being England's most well-known planning policy, it is also widely misunderstood, with just 28% of respondents saying that they know a 'great deal' about it whilst 72% say that they know 'just a little/(have) never heard of/ heard of but know nothing' about it (Table 7.2). This demonstrates the qualitative point made by planners about the public not understanding the policy and makes a strong case for more planning education for the public.

There are even more significant differences among different ages and class groups with a 33% difference between those aged 65+ and 25–34-year-old

Table 7.1 Responses to the statement that the Green Belt should be retained and not built on

	Agree
Gender	
Male	63%
Female	66%
% Difference	3%
Social Class	
AB	72%
DE	54%
% Difference	18%
Age	
25–34	57%
65+	73%
% Difference	16%
Property Type	
Owner-occupier	72%
Mortgage	65%
Socially rented	58%
Privately rented	57%
% Difference (Highest-lowest)	15%
Geographical Area	
South	72%
Midlands	61%
% Difference	11%
Qualification	
No Formal	57%
Degree	68%
% Difference	11%
Type of Area	
Semi-rural	83%
Urban	62%
Rural	72%
% Difference	21%
Total	64%

Source: Constructed with data from Ipsos MORI (2015, pp. 4–8).

claiming to know a great deal (41–8%) and little about it (92–59%) about it. This lack of knowledge among 25–34-year-old again highlights the need for more planning education with knowledge of it greatest among the 'baby-boomers'[3]. Additionally, there was a large disparity (26%) among social class *suggesting that it is particularly well-known and well-supported* among higher social groups, perhaps as preserving semi-rural 'residential exclusivity' (Short *et al*, 1987, p. 36), where these people are most likely to live (83% of respondents in 'rural-urban' areas

Table 7.2 How much do you know about the Green Belt?

	A great deal:	Just a little/never heard of/heard of but know nothing:
Gender		
Male	5%	23%
Female	6%	28%
% Difference	1%	5%
Social Class		
AB	42%	58%
DE	16%	84%
% Difference	26%	26%
Age		
25–34	41%	92%
65+	8%	59%
% Difference	33%	33%
Property Type		
Owner-occupier	39%	61%
Mortgage	35%	66%
Socially rented	21%	79%
Privately rented	13%	87%
% Difference (Highest-lowest)	26%	26%
Geographical Area		
South	68%	70%
Midlands	70%	68%
% Difference	2%	2%
Qualification		
GCSE Level	20%	80%
Degree	37%	63%
% Difference	17%	17%
Type of Area		
Semi-rural	27%	56%
Urban	44%	73%
Rural	33%	67%
% Difference	17%	17%
Income		
Above £25,000	8%	64%
Below £24,999	4%	77%
% Difference	4%	13%
Average % difference	28%	71%

Source: Constructed with data from Ipsos MORI (2015, pp. 1–3).

supported it compared to 62% in urban areas). Knowledge was significantly higher (26%) among owner-occupiers than those privately renting and there was a knowledge gap (17%) based on educational qualification and income (13%). Nonetheless, whilst there is *a* class and economic dimension in support for, and knowledge of, the policy, it remains widely supported but poorly understood among a *range* of social and economic groups.

The detailed breakdown of data is unavailable for the 2005 questionnaire although the amount of people saying that they knew 'little to nothing' about Green Belt was almost identical at 72% (Iposs MORI, 2005). The proportion of people *disagreeing* that it should be protected increased from 5% to 17% from 2005 to 2015 suggesting that there is greater scepticism of the policy now.

SOCIAL ATTITUDES SURVEY – UNDERSTANDING SOCIETY

This a longitudinal dataset although identical questions were not asked annually on the policy and housing. Notwithstanding the critique of CPRE's questionnaire being partially self-selecting (Lane, 2019), the proportion of people agreeing with the statement '*Keep Green Belt, do not build there*', 76.2% (1997), 72.7% (1998) and 79.8% (1999), was similar to the 84% in CPRE's 2005 Questionnaire. If these slight differences are down to surveying a different group of people each year, it *suggests* the key trigger for decline in support for Green Belt is increasing awareness of the housing crisis since 2005 rather than a sampling bias.

COMRES (2018) QUESTIONNAIRE

This is a useful poll and it appears that attitudes towards the Green Belt have become more critical even since CPRE's 2015 Questionnaire (Table 7.3).

This deals with views in *principle* rather than a specific development in one's *local* area, which, drawing on place attachment theory (Devine-Wright, 2009), often motivates opposition and the questionnaire is people's views specifically on land which is *least attractive*. Nevertheless, given the support that it is popularly perceived to have, there was only a 10% difference overall between those agreeing (32%) and disagreeing (42%) with the statement. This does not fundamentally undermine the argument of it being widely supported as a *matter of principle* but suggests that a significant number of people are willing to see *particular parts of it, especially the least attractive land, released in particular circumstances*. There was also a significant proportion who 'do not know' (26%), but compared to CPRE's questionnaire, the differences between social groups are not so significant. Particularly surprising is that, although more of 65+ age group (45%) than the 18–24 group (35%) disagree with the statement, the same margin (10%) agree with the statement (40% among 65+ compared to 30% among 18–24) suggesting that there is greater knowledge now of the housing crisis across generations.

The next question focused on TOD in the Green Belt (Table 7.4):

There was only a 10% gap between agree (31%) and disagree (41%), again suggesting that a significant proportion of people are willing to see *particular parts of the Green Belt released where there is supporting infrastructure*. Similarly, there were not significant differences across social groups although the amount of people

Table 7.3 Green Belt should be loosened for least attractive land

	Agree	Disagree	Don't know	Difference Between Agree and Disagree
Total	32%	42%	26%	10%
Age				
18–24	30%	35%	35%	5%
65+	40%	45%	15%	5%
% Difference	10%	10%	20%	0%
Gender				
Male	35%	44%	21%	9%
Female	30%	40%	30%	10%
% Difference	5%	4%	9%	1%
Social Grade				
AB	36%	43%	22%	7%
DE	29%	39%	32%	10%
% Difference	7%	4%	10%	3%
General Region				
Midlands	32%	43%	25%	11%
North	32%	41%	25%	9%
South	31%	42%	27%	11%
% Difference	1%	2%	2%	2%
Specific Region				
South East	27%	49%	24%	22%
West Midlands	31%	42%	28%	11%
% Difference	4%	7%	4%	11%
Political Party				
Labour	46%	41%	29%	5%
Conservative	38%	46%	16%	8%
% Difference	8%	5%	13%	3%
Type of Area				
Urban	33%	41%	27%	8%
Rural	31%	47%	22%	16%
% Difference	2%	6%	5%	8%

Source: Constructed with data from the ComRes Questionnaire (2018, pp. 5, 12–15).

agreeing in the AB category (38%) being significantly more than the DE category (25%) was surprising given the popular perception of the policy as supported by the wealthy (although both groups have a significant segment who 'do not know' 24% (AB) and (25% (DE)). The same point could be made regarding slightly more Conservative voters agreeing (35%) than Labour ones (30%) with the Conservative Party often associated with protecting the policy (Goode, 2022A). Nonetheless, proportionately more Conservative voters (45%) disagreed with the statement compared to Labour ones (40%) whilst there was a significant proportion of 'do not knows' 30% (Labour) and 20% (Conservative). This suggests that, whilst there is still widespread support for the Green Belt, this support is *decreasing* due to the deepening housing crisis.

Table 7.4 Green Belt restrictions should be relaxed around train stations

	Agree	Disagree	Don't know	Difference Between Agree and Disagree
Total	31%	41%	29%	10%
Age				
18–24	31%	33%	36%	2%
65+	37%	46%	17%	9%
% Difference	6%	13%	19%	7%
Gender				
Male	35%	27%	24%	8%
Female	27%	40%	33%	13%
% Difference	8%	13%	9%	5%
Social Grade				
AB	38%	38%	24%	0%
DE	25%	38%	25%	13%
% Difference	13%	0%	1%	13%
General Region				
Midlands	32%	39%	29%	7%
North	29%	42%	30%	13%
South	31%	42%	27%	11%
% Difference	3%	3%	3%	0%
Specific Region				
South East	27%	49%	24%	22%
West Midlands	31%	42%	28%	11%
% Difference	4%	7%	4%	3%
Political Party				
Labour	30%	45%	30%	15%
Conservative	35%	40%	20%	5%
% Difference	5%	5%	10%	0%
Type of Area				
Urban	30%	39%	30%	9%
Rural	32%	46%	22%	14%
% Difference	2%	7%	8%	5%

Source: Constructed with data from the ComRes Questionnaire (ComRes, 2018, p. 5, 14–15).

Other Questionnaires

BROADWAY MALYAN

This questionnaire, carried out by YouGov (2015, p. 3), helpfully asked a 'principle' question on whether the public supported building in the Green Belt (Table 7.5).

As this questionnaire was also conducted in 2015, there were similar results to CPRE's one with widespread public support for the policy, 67% opposition compared to 17% support, compared to brownfield development (83% support and 3% opposition). Levels of support are slightly higher (10%) in support of greenfield than Green Belt development and *significantly* lower in opposition (19%). This suggests that the Green Belt as a *principle* commands more support than greenfield

Table 7.5 To what extent would you support or oppose
new housing being built on Green Belt land?

	Support	*Oppose*
Total	17%	67%
Age		
25–39	20%	61%
60+	15%	75%
% Difference	5%	14%
Gender		
Male	20%	63%
Female	14%	70%
% Difference	6%	7%
Social Grade		
ABC1	17%	70%
C2DE	17%	63%
% Difference	0%	7%
Specific Region		
South East	15%	71%
West Midlands	14%	72%
London	27%	55%
% Difference	13%	17%
Political Party		
Labour	21%	60%
Conservative	16%	74%
% Difference	5%	14%

From YouGov (2015, p. 3).

land despite both being largely about the loss of countryside. This questionnaire had a bigger sample (4510), compared to CPRE's (845), with some popular perceptions of social differences emerging more clearly. Most prominently, 14% more Conservative voters supported the Green Belt than Labour ones resonating with the popular image of the Conservatives as the 'natural' supporters of the policy (Inch *et al*, 2020). Social differences between classes are not pronounced (7%) but classes have been grouped together (ABC1 and C2DE are compared to AB and DE in other questionnaires). Generational differences stand out more with a 14% difference between opposition to Green Belt development among those 60+ and between 25 and 39. Information on property tenure was not gathered, but again, the trend is that the Green Belt's supporters *tend* to be older, Conservative voters, but that it still retains *widespread public support*[4]. The South East and West Midlands are almost identical in their support of it reflecting the South East having the highest levels of homeownership and development pressure alongside the WMGB's popularity.

HOME OWNERS ALLIANCE (HOA) (2015)

The HOA (2015, p. 11) commissioned a YouGov survey of homeowner attitudes (2184 adults) to various housing and planning policies before the 2015 Election permitting useful cross-comparison with the CPRE and Broadway Malyan

Table 7.6 Support for building on Green Belt land with little environmental or amenity value and Garden Cities

	Green Belt			Garden Cities	
	Support	Oppose	Neither agree/ disagree	Support	Oppose
Total	27%	56%	13%	50%	
Aspiring Homeowners		37%		59%	
Regions					
London	39%			57%	
West Midlands	19%			42%	
Gender					
Male	32%			55%	
Female	22%			45%	

Source: From the HOA (2015, pp. 14–17).

Questionnaires (also conducted in 2015). Although the HOA (2015) runs an annual survey, questions on planning policies, including the Green Belt, were a one-off, so the data does not permit longitudinal analysis and the raw data is unavailable, but it has been helpfully segmented (see Table 7.6).

Again, as the question is specifically on *poor quality land* in the Green Belt rather than as a *principle*, there is less opposition (56%) and more support (27%) for releasing land (compared to 67% opposition and 17% support in the Broadway Malyan Questionnaire). Opposition falls to 37% among prospective homeowners whilst support for Green Belt development is highest in London (39%), probably due to its housing crisis. This further supports the conclusion that, whilst the policy commands significant support in *principle*, in *practice* a significant proportion of people are willing to see a *limited* release of land *under particular circumstances* to help address the housing crisis. Surprisingly, the West Midlands was the region with the most opposition to Green Belt release, again reflecting the way that the WMGB is popular as containing the 'industrial' conurbation (Goode, 2022B). Gender was not expected to be a significant factor, but this questionnaire shows a 10% difference in support for the policy. Although most people had clear views on the policy, a minority of people (13%) neither agreed/disagreed with the statement. Finally, support for garden cities and new towns, 50% compared to 27%, is higher than releasing Green Belt land highlighting another potential solution to the housing crisis (although questions remain as to the *location* of new towns) (HOA, 2015, p. 6).

Green Belt release and building new homes remain unpopular compared to demand side policies with 80% supporting a policy of marketing homes to UK buyers first, 70% higher council tax on unoccupied homes and 65% extending Help to Buy. Nonetheless, 70% acknowledged that the availability of housing *is* an issue showing widespread societal concern about the housing crisis (HOA, 2015, p. 6–8).

FEDERATION OF MASTER BUILDERS (FMB) (2018)

This questionnaire of 2000 homeowners explored potential solutions to the housing crisis and, again, found the societal preference for demand side and smaller-
scale policies, with 33% supporting co-living developments and 31% micro homes
(FMB, 2018, p. 1). It was similar to other 'principle' questions, like YouGov's
(2015) one, with only 17% supporting Green Belt development. However, this
survey was of homeowners and a choice of policy 'options' were given.

HATTON PARISH PLAN (2013)

This questionnaire of an affluent rural parish on the edge of Meriden Gap showed
that 60% of respondents opposed development in the countryside whilst 70%
argued that development in the Green Belt should be resisted (although 25%
thought that the boundaries should be reviewed) (Hatton Parish Plan Steering
Group, 2013, pp. 12–13). This shows the inseparability of popular support for the
countryside and Green Belt although interestingly, *even in an affluent parish*, a
significant minority take a pragmatic view of the policy.

Popular Attitudes towards Housing Development

Popular attitudes on housebuilding more broadly are explored through the quantitative datasets to further triangulate and generalise the study.

Social Attitudes Survey – Understanding Society

SOCIAL ATTITUDES SURVEYS SINCE 1990S

This dataset is particularly useful given its longitudinal nature although questions
on housing are not asked annually neither are the same questions asked – questions
were asked in 1997–1999, 2010, 2013–2014 and 2016–2018 but not in the other
years (Park *et al*, 2012). Whilst this data is over 20-years-old, it provides a fascinating historical snapshot of societal attitudes with a significant number of people recognising a potential housing crisis although there were high(er) levels of popular
opposition to new housebuilding. There was still a prevailing preference towards
suburban development being at the beginning of New Labour's Urban Renaissance
(Hall, 2002).

SOCIAL ATTITUDES SURVEYS SINCE 2010

There has been a significant reduction in opposition to development, probably as
public awareness and the political importance of the housing crisis has increased
(see Figure 7.1). Although opposition has levelled out recently this has been a significant change in attitudes in a short timeframe (Gallent, 2019)[5].

Attitudes Towards Housebuilding

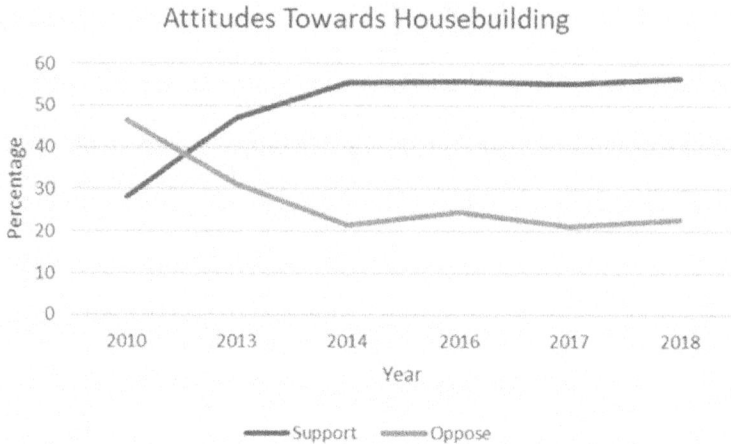

Figure 7.1 Support and opposition to more homes being built in one's local area (Constructed with data from Figure 1.1, Ministry of Housing Dataset – Chapter 1: Figures and Annex Tables (2018)).

The *likelihood* of opposing development has also reduced again reflecting greater societal concern about housing (MHCLG, 2018).

SOCIAL ATTITUDES SURVEY, 2018

The statistics on attitudes towards housebuilding are very illuminating (MHCLG, 2018):

Profile of Those Opposing Development　The results clearly show that those *most likely to oppose development generally* tend to be owner-occupiers, older, wealthy and live in rural/semi-rural areas. People aged 46–55 have the highest *likelihood* of opposing development may be because they are likely to be completing mortgage payments. This general profile resonates with the broader literature on opposition to development (Sturzaker and Shucksmith, 2011; Wills, 2016). Nevertheless, opposition to housebuilding is clearly wider than *solely* Green Belt reasons although the data does not highlight the *group(s) specifically* campaigning on the policy.

Strategies for Opposing Development　There is a mixture of traditional and more modern oppositional techniques (see Chapter 8). Signing a petition is the most popular, probably because it can be done online and is the easiest way for campaigners to garner support but it is often viewed sceptically by planners, especially when a questionnaire does not include the signees postcode. Interestingly, attending a public meeting remained popular, despite being time-consuming, perhaps relating to the sociability and drama involved. Submitting a formal objection about development was more popular than writing to one's local councillor. Finally, given its potentially time-consuming nature, joining an action group was the least popular

What would make you support new development?

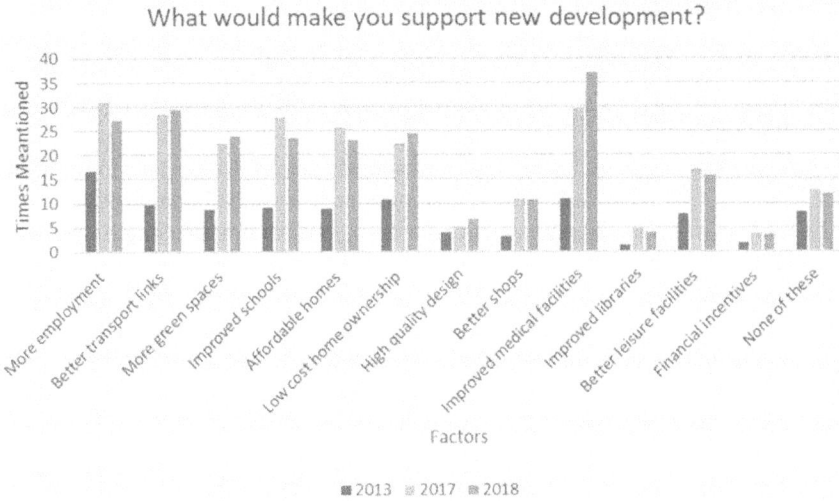

Figure 7.2 Advantages to local residents to make them support homes being built in the local area (Constructed from: Annex Table 1.9, Ministry of Housing Dataset – Chapter 1: Figures and Annex Tables (2018)).

technique. There is a significant *gap* between *attitudes,* in terms of *wanting* to oppose development, and *actions,* so a relatively small group of people *are actively involved in campaigning.*

Motivations for Opposing Development and Improving the Public Acceptability of Housing   Unfortunately, this dataset does not contain the *direct* motivations for people opposing development although this can be partly derived from the factors which people identify as lessening their opposition[6] (see Figure 7.2). Factors increasing in importance recently relate to infrastructure and facilities. Medical facilities are now the most important factor by a significant margin, probably more so since the COVID-19 Pandemic, although better transport links, schools and greenspace all feature as important. Employment opportunities also appear, and affordable housing has been increasing in importance reflecting greater societal concern about housing. A number of respondents (12% in 2018), said that none of the above factors would make them more likely to support development suggesting that there is a (minority) of people who oppose development in every circumstance thus resonating with the 'normatives' described by Mace (2018, p. 1).

Indeed, financial incentives are not a significant factor as demonstrated in the Government dropping the proposal of direct financial incentives to homeowners to reduce opposition to local development (due to the lack of popular support) (Inch *et al*, 2020)[7]. Conversely, the expense of newly built housing *was* a big concern (with CPRE (2018) frequently highlighting the cost of new housing built in the Green Belt)[8]. This could be conceptualised as a cynical attempt by campaigners to discredit new housebuilding as too expensive to serve their underlying objective of preventing development but, with this large dataset, it encompasses a larger group than *just* campaigners so is reflective of wider societal concerns (Amati, 2007). Moreover, it perhaps suggests that most people tacitly accept new housebuilding does *not* lower house prices but still recognise the need for new housebuilding.

Nonetheless, there are higher levels of support for affordable housing – in the 2016 survey, 72% said that they would support more housebuilding locally *if it was affordable* whilst 53% mentioned social/housing association housing as the most needed tenure locally compared to 11% referring to rent from private landlords (NatCen Social Research, 2017). This reflects the broader softening in societal attitudes towards social housing with 71.6% of people saying that they support financial assistance to people on low incomes renting (MHCLG, 2019; NatCen Social Research, 2017). Overall, these results show that opposition to housebuilding is not *solely* about material, economic concerns, as the house price hypothesis presumes, but includes a mixture of place attachment and principled concerns (Anton and Lawrence, 2014; Inch *et al*, 2020). Although focused on a specific Urban Growth Boundary (Green Belt), these results have wider relevance to the literature with the shared characteristics of opposition to housebuilding in relation to densification, rural development and affordable housing.

Other Aspects of Housing and the Housing Crisis   The overwhelming *cultural preference* towards homeownership has marginally increased in recent years with the deepening housing crisis and problems associated with private renting (Christophers, 2018, 2019). In 2013, most people agreed that there was a shortage of new homes *nationally* (81.7%) and in their *locality* (70.9%) with 83.4% agreeing that it was easier to buy a house 20 years ago[9]. Again, there is more public support for *demand side policies* with 36% of people mentioning financial assistance to first-time buyers, such as Help to Buy, compared to 3.6% who mentioned making it easier for housebuilders to get planning permission. The central challenge for planners is still that, whilst people now agree that there is a crisis *in principle*, specific, local housing developments are often very unpopular and the Green Belt still *largely* retains its popularity (Lane, 2019). Additionally, 56.3% agreed with the statement that 'housing will remain unaffordable in my area, even if there is new housebuilding' showing widespread public scepticism about new housebuilding so affordability needs to be a key factor in addressing the housing crisis.

Broadway Malyan/YouGov Questionnaire (2015)

This questionnaire asked some helpful questions (Table 7.7).

This shows the extent of public concern about housing, especially for first-time buyers, and there is recognition of more homes needing to be built locally (67% acceptance). Whilst this shows the contradictory nature of polling data on popular attitudes towards housebuilding, it again calls into question the primacy of the house price hypothesis.

Table 7.7 Broadway Malyan Questionnaire results

How easy or difficult is it for the following groups to buy and rent in your local area?

	Easy	*Difficult*
Local People	21%	51%
First-time buyers	11%	68%
Should the number of homes built increase or decrease in your local area?		

	Increase	Decrease
	67%	8%

Constructed with data from YouGov (2015, p. 2).

The ComRes and YouGov Polls (2018)

Again, this reveals extensive societal concern about housing and a desire to build *moderate* levels of new housing (68% nationally/59% locally) and bring house prices down *moderately* (59% nationally/52% locally) (Table 7.8). Once more, this calls into question the house price hypothesis (Inch *et al*, 2020). Nonetheless, there are significantly less levels of support for *large-scale* housebuilding nationally than locally, a 17% differential, and less for bringing house prices down moderately (a 7% differential). To some extent, there are 'NIMBY' attitudes, or an acknowledgement of the need for *more* housebuilding nationally but not so much locally (DeVerteuil, 2013) but the majority of people in this questionnaire, nationally *and* locally, *do* support a moderate level of housebuilding *and* a moderate reduction in house prices (Table 7.9).

The ComRes poll had a large sample (2036) and showed widespread societal concern about housing, with 63% compared to 19% acknowledging that house prices are too high locally and 61% compared to 15% recommending that a newly married couple buying a house over renting (Table 7.9). Whilst there are still significant levels of opposition, there is overall support for new housebuilding in *local* areas (48%) and for some land to be used around towns and cities (47%) for development, showing that most people support housebuilding *in certain circumstances*. There is a minority, again resonating with Mace's (2018, p. 15) characterisation of 'normatives', who oppose development in most circumstances. Even when asked *specifically* about affordable housing, more people (45% compared to 31%) *disagreed* that protecting the countryside prevented affordable housing from being built showing widespread popular support for the countryside.

A majority (61%) agreed that higher quality housing would lessen their opposition which, although higher than the other questionnaires, may be explained by the fact that this questionnaire asked *specifically* about design rather than as a list of options. 63% agreed that more community benefits would make them more likely to support development, showing their importance, although the questionnaire did not ask specifically about *what* these benefits would be. Finally, the lack of public knowledge about planning is obviously wider than the Green Belt, with 22%

Table 7.8 YouGov Questionnaire

	Support	Oppose	Difference in Support Between National/Local
National Level To what extent would you support the Government attempting to…			
Build a large number of new homes	60%	30%	17%
Build a moderate number of new homes	68%	21%	12%
Build a small number of new homes	41%	46%	−11%
Stop any new homes from being built	9%	79%	-6%
Local Area To what extent would you support the Government attempting to…			
Build a large number of new homes	43%	44%	
Build a moderate number of new homes	56%	32%	
Build a small number of new homes	52%	35%	
Stop any new homes from being built	15%	71%	
National Level To what extent would you support the Government attempting to…			
Bring house prices down a lot	52%	33%	3%
Bring house prices down a moderate amount	59%	25%	7%
Bring house prices down a little	52%	32%	3%
Keep house prices about the same	31%	51%	−5%
Push house prices up a little	13%	70%	−4%
Push house prices up a moderate amount	10%	74%	−1%
Push house prices up a lot	6%	79%	−2%
Local Level To what extent would you support the Government attempting to…			
Bring house prices down a lot	49%	36%	
Bring house prices down a moderate amount	52%	31%	
Bring house prices down a little	49%	34%	
Keep house prices about the same	36%	45%	
Push house prices up a little	17%	66%	
Push house prices up a moderate amount	11%	71%	
Push house prices up a lot	8%	75%	
National Level Which option would you prefer?			
More houses are built and house prices go down	63%		8%
House prices remain the same and no houses are built	18%		−10%
Local Level Which option would you prefer?			
More houses are built and house prices go down	55%		
House prices remain the same and no houses are built	28%		

Data from YouGov (2018, pp. 1–6).

Table 7.9 The ComRes Poll (2018)

	Agree	Disagree	Don't Know
I would support more homes being built in local area	48%	33%	19%
While most of the land around England's towns and cities should be protected, some should be used for development	47%	32%	21%
Protecting the countryside around England's large towns and cities prevents affordable housing from being built	31%	45%	23%
Property prices in my area have become too high	63%	19%	19%
I would be more likely to support development if there were more community benefits	63%	16%	21%
I would be more likely to support development if the quality of old buildings matched the new	59%	18%	24%
I would advise a newly married couple to buy a house rather than rent	61%	15%	24%
Average			22%

Source: Information from ComRes (2018, p. 5, 22–23).

answering on average 'don't know' in response to statements highlighting the need for greater public planning education.

Churchill Home Insurance (CHI) Poll (2019)

A questionnaire exploring motivations and NIMBYism was conducted by Opinium (commissioned by CHI) (all data from Morris, 2019, p. 1). Its usefulness is limited as the raw data is not publicly available, so it cannot be analysed according to social groups etc., and it examines domestic planning applications, such as home extensions, alongside larger housing developments. Nevertheless, there are some interesting findings, including that there were 1.9 million objections between 2017 and 2019[10]. The second joint highest reason for objection was the loss of view (50%), the third was the impact of development on property prices (43%) and the fourth was the impact of construction works (39%). Whilst this clearly incorporates those who are *directly* affected by development, it shows that emotional attachment to place and loss of view is an important motivation (Anton and Lawrence, 2014). An article in *Property Investor Today* (Lane, 2019, p. 2) argued that the poll showed that NIMBYism was 'on the rise' again, although 54% of respondents reported feeling powerless with councils not listening to them.

Conclusions on Quantitative Data on the Green Belt and the Housing Crisis

Although *campaigns* against development generally are dominated by wealthy, older people with high levels of social capital, there is insufficient *quantitative* data to firmly demonstrate that these campaigns *materially affect planning outcomes* and that the planning system *itself* leads to *significant* power imbalances (see Chapter 8) (Inch *et al*, 2020). The neo-Marxian framework has limited utility nationally as this chapter's findings have problematised the house price hypothesis whilst only a *minority* of homeowners are actively involved in campaigns (Matthews *et al*, 2015).

The Green Belt appears to command widespread public support for emotional, reasons of principle, as popularly perceived as protecting England's 'green and pleasant' land, alongside place attachment (Mace, 2018). However, as the *status quo*, support for it is arguably more muted than if it was popularly perceived to be 'under attack' with, say, the Government proposing its abolition (Amati and Taylor, 2010, p. 143). Nevertheless, the data suggests that there are *certain* circumstances in which people support Green Belt release for housing thereby suggesting policy possibilities of improving the public acceptability of new housebuilding. More broadly, these findings confirm Cherry's (1982, pp. 116–117) wider critique of Marxian theory: 'The reality is much more muddied…the enforced superimposition of the simple on the complex is just not convincing'.

Qualitative Data

Exploring the data on what campaigners say about themselves is important to understand how and why specific, regional Green Belt campaigns can often be so passionate, although this section is largely on the views of *professional* campaigners[11]. As with planners, campaigners and politicians may have vested interests, i.e. protecting their local area, ensuring political support etc., and often represent a particular demographic, but they still play a vital role in planning[12]. To try to capture the views of everyday campaigners, three active groups were engaged with (Save Stourbridge Green Belt (SSGB), South Solihull Community Group (SSCG) and Project Fields (PF) (see Chapter 2; Figure 2.3)).

Matters of Principle, Place and Property

A key feature of professional campaigners was their desire to demonstrate that they were not 'NIMBYs', recognised the housing crisis and understood how planning and the Green Belt 'worked', like the distinction between greenfield and Green Belt (all seven had planning backgrounds). Although property values were mentioned as *a* factor in opposition to development, campaigners stressed other factors, particularly 'legitimate' planning reasons, especially concerns about infrastructure and the Green Belt's strategic purpose. The views of professional campaigners can also be grouped into three broad themes: 'principle', 'place' and 'property' arguments. This finding regarding the professionalisation of campaigning reflects Roy's (2011, p. 591) work on women's movements in India. Similar points were made by everyday campaigners to professional campaigners although, unsurprisingly, they described their concerns in more direct, everyday language but the substance of their views was similar to 'professional' campaigners. Politicians tended to use more sophisticated language than everyday campaigners although, again, focused on more emotional arguments (see Chapter 8).

Figure 2.3 (repeated): Approximate locations in which the campaign groups are operating (adapted from: https://bit.ly/3pQUZdv).

Matters of Principle: Support for the Green Belt and Countryside

Firstly, campaigners underlined the importance of general, popular planning principles as, in many ways, a popular battle rages over hearts and minds as to which group, campaigners or developers, represent the public interest (see Chapter 8; Bradley, 2019A). Although it was difficult disentangling whether support for the Green Belt arose from support for *the policy* or broader countryside, generally professional campaigners supported the principle whilst everyday campaigners largely wanted to prevent development in the countryside with the Green Belt employed as an oppositional strategy or 'tool'. For example, regarding the Green Belt:

> I have never seen policies as successful as the Green Belt for a small country so I very strongly support Green Belt. **I think the name is an awkward one because people say – 'Well, you know, it's Green Belt, it's about greenfields (and it's about prevention of coalescence, preventing sprawl and the openness of the countryside). But the Green Belt policy is primarily a good one...**the West Midlands is a very diverse region...you need a firm commitment to protect the Green Belt and ensure that, **before you allow greenfields and Green Belt to go, you have tackled as many brownfields as you can**.
>
> (Local councillor, Solihull (individual views))

Then, in relation to the principle of protecting the countryside:

> The balance between things will differ from area to area...Green Belt pro-
> vides what most people see as a high degree of certainty that an area that
> is currently open and free from development will remain so. **I don't think
> there are very many people, laypeople say, who understand the nuances
> of exceptional circumstances and very special circumstances...All that
> is just to them a detail of Government policy that they needn't concern
> themselves with. They are interested in the Green Belt at a much more
> basic level and 'Is that greenfield over there going to stay a greenfield?'**
> I like the view and chance to walk across that footpath over the area because
> it is a beautiful area and I see lambs and cattle and all the rest of it and those
> are the things that matter to people.
>
> (Planner (1), CPRE West Midlands)

Often, support for the Green Belt and countryside was interwoven as was apparent in the SSGB responses:

> Green Belt is countryside – the two are not exclusive.

> Whether the land proposed for development is Green Belt or open coun-
> tryside **is largely immaterial from the point of view of the effects of
> development**.

The importance of the countryside in the WMGB is poignant as resonating with
the quantitative data and the historic antipathy of the shires towards encroachment
by the 'industrial' conurbation (Goode, 2022B).

Place: Protecting One's Local Area

The concrete effects of development upon one's local area, especially infrastructure and services, featured prominently alongside the interconnected, more abstract
fear of change, especially about an area's 'character' changing. Again, this featured
strongly in SSGB's responses. For example:

> There is a lack of vision in the building on greenspaces, take the scheme out-
> side Hagley. No thought was given to the infrastructure, so there is now con-
> gestion, pressure on schools, doctors, local shops and (the) hospital. Profit
> comes first with little attention being paid to the afterwards.

Congestion and infrastructure were also highlighted by the PF campaigner and
these points were raised more generally by professional campaigners and politicians. The more subtle concerns about the effects of development upon an area's
(often) exclusive, semi-rural 'character' were not so clearly voiced but could be
detected, especially in the West Midlands:

If you have new housing…**you will fundamentally change the character of this area and it topples over. So, you no longer have got an attractive area to attract investment…actually, you need, what I call some 'Rolls Royce areas'**…The premium sites should be premium sites. They should only be able to provide what you cannot provide elsewhere and make this area special to be able to attract that sort of investment. And the great danger is that it will become an 'anywhere place'. So, sensible planning would try to manage this to make sure it is more sustainable. [Housing] is proposed on a very large scale in Knowle…if you look at the car parking and road situation it is difficult now already…it's dangerous etc. and you are changing the character of the place, quite fundamentally…you are extending urban development into the rural area, rural scene…it will be very damaging

(Retired Strategic Planner, West Midlands)

These quotes suggest that campaigners opposition is strongly shaped by the local environment and environmental psychology with high levels of place attachment (Bailey *et al*, 2016; Cowell *et al*, 2011; van den Nouwelant *et al*, 2015).

The Property Argument

In many ways, an area's semi-rural character is bound up with property prices (Rydin, 1985) but, as with private sector planners, campaigners did not locate property prices as the *primary* factor in campaigner and wider popular support of the Green Belt:

A lot of urban people are concerned about the loss of Green Belt…the people who object to housing nearby to them usually feel they are protecting their own countryside. Some of it maybe the house prices but I get the impression, quite often, that they have moved there…and their main concern is loss of their countryside, amenities and values.

(Policy and Campaign Advisor, West Midlands)

Notwithstanding the challenges of establishing motivations for opposing development, the evidence *suggests* that property prices are not the *primary* motivation and popular planning principles and place attachment appear more important. Again, this resonates with the cultural geography and place attachment literature more than neo-Marxian theory (Davison *et al*, 2017; Devine-Wright, 2009).

Localist or Strategic Views of the Green Belt?

This quote demonstrates the strategic concerns of professional campaigners' (see also Chapter 9):

A strategic approach…I can only speak from this region – we would do very well with that. The trouble now is we don't have that management…in terms of governance. Is there sufficient cooperation in strategic planning which enables authorities to manage and allocate the [BDP] shortfall? Well, no…it has engaged in fudging and fixing. In the past…the counties [were] doing Structure Plans and all getting together with the Government Office and either the county or city authorities. We have now got…a situation where one authority leads in saying to another – Will you take 2,000 houses? It is a sort of political fix. There was an element of that, probably 40/50 years ago, but it has now become really pretty obvious that it has not been done by strategic planning.

(Planner (2), CPRE West Midlands)

However, more local, parochial concern and ownership of the policy became apparent among everyday campaigners posing challenges regarding the rebuilding of strategic planning. Again, the antipathy of semi-rural areas in the WMGB to the 'industrial' conurbation emerged strongly:

(Langley) is being forced on Sutton by the **Labour Council!**…The truth is this is completely political -the Conservatives would never dream of doing this! They have far too many votes in Sutton!…Langley is a leafy, nice part of Birmingham…near the countryside.

(Campaigner, PF)

Nonetheless, the Conservative politician from Sutton Coldfield had become more favourable towards strategic planning when interviewed for the second time and he envisaged a 'Black Country Garden Village' accommodating 40,000 homes. The PF campaigner argued that the scale of the West Midlands' housing need meant that planners needed to 'start again' through reviving the 'new towns' policy with settlements far enough away from Birmingham to be self-sufficient such as Tamworth (Walker, 2016). Whilst these could be conceptualised as NIMBY strategies to legitimatise opposition to Green Belt release and move housing 'need' elsewhere, they show that strategic consideration is sometimes given by campaigners and politicians. Nevertheless, antipathy towards strategic planning in the West Midlands became apparent amongst everyday campaigners as this SSGB campaigner quote make clear:

A regional approach to meeting housing needs would be unfair…Councils and residents with Green Belt land would be penalised by having to meet the targets of those boroughs that had not met their brownfield sites targets as well as their own.

Further research is required to explore how far strategic views of the Green Belt are held by everyday campaigners in different regions and campaign groups and whether opposition to Green Belt development varies significantly in character to greenfield development. More research is also needed into campaigner attitudes to Urban Growth Boundaries in other countries, such as the Golden Horseshoe in Ontario. However, there is a divergence in this data between everyday and professional campaigners in the West Midlands. Chapter 9 establishes certain safeguards for strategic planning regarding the Green Belt to build public confidence.

Interactions Between Planners and Campaigners

As outlined in Chapter 5, tension between planners and campaigners is partly inevitable given that they have different, often competing interests and priorities, whilst planning systems in many countries are inherently conflictual and confrontational as intimately bound up with place, especially people's lived experiences (Kiernan, 1983). However, the *depth* of mutual distrust between campaigners and planners was troubling because an effective and well-functioning planning system ultimately relies upon public confidence that planners act with professional integrity (Goode, 2022A; Parker and Street, 2015; Raynsford Review, 2018B). Whilst this may be an idealised 'compact' or 'contract' between planners and the public (Raynsford Review, 2018B, p. 117), how planning is popularly *perceived* is very important (Goode, 2022B). In terms of broader issues explaining the public's lack of trust in planners, which are beyond this study's scope, a common theme emerged among planners of the wider distrust of 'experts' and planning is often targeted for things that it is not directly responsible for, i.e. school and GP places or traffic congestion. This has been exacerbated by the localism agenda and two-tier system of local Government compared to other countries which have more strategic and/or federal forms of governance (see Chapters 6/9). As LPA Director (West Midlands) reflected:

> **The planning system is an easy target but it is a mechanism for decision-making**...this is **where we need to be more joined up**...[it] needs to link in more carefully with the healthcare planning system...there is a necessity for us to be far more strategic in our approach and bring these bodies and mechanisms closer together so that **you don't have a situation where you are seeing housing growth but you are not seeing the commensurate service facilities. That is where planning gets criticised – we are very good at planning for housing and jobs but not very good at planning and delivering the essential service and infrastructure that are needed to support the community**.

This returns to the lack of control that people feel they have over their lives and local places with planning often blamed by people in their fear of change. Of many quotes, the planner from the TCPA probably spoke most passionately:

> House prices are just a thing. It is a metaphor for how they manifest their actual concern... So, there is an issue about people's needs... the transport, social and medical infrastructure is not there to support that development...[people of] the deregulation kind of don't understand that all of these concerns are actually legitimate...where we ended with Raynsford[13] is a fundamental breach of trust. The 'social contract' in planning that existed...[during] the post-war era was one where people had a voice in

planning and where they elected people in planning who they thought had a voice in planning...[Now] I don't think they trust planning or planners. They certainly think it is a development-led approach. So, what are you going to do without core trust? It is a very difficult process...we went so far in Raynsford to talk about community rights because, what we were trying to do was to say to those in the development sector and the planning profession who don't listen to this because it is inconvenient – 'you are going to have to work very hard to bring trust back into the system'.

Planners also highlighted how the frequency of ideological 'attacks' on planning alongside austerity, deregulation and the stripping away of its responsibilities to an (essentially) regulatory service, has reduced its effectiveness in delivering successful place-making. Moreover, these factors have reduced its ability to bring together different bodies to coordinate things, such as the infrastructure and healthcare needed, alongside development as is done more effectively in Sweden and the Netherlands, for example (Adams and Watkins, 2014; Kenny, 2019A, 2019B).

Rebuilding Trust in Planning: Education, Engagement and A National/Strategic Green Belt 'Conservation'

This section develops recommendations articulated by planners regarding community education and engagement in planning which sit alongside recommendations in Chapter 6 regarding more effective funding of local services and in Chapter 9 about the need for strategic planning. Based upon the data, there is a strong case for increasing public planning education before engagement, especially on the Green Belt as the most well-known but also misunderstood planning policy. Additionally, consultation needs to include a wider range of people. These two ideas are woven together in the recommendation of a national Green Belt debate.

Engaging a Wider Range of People and Planning Education

Private Sector Planning Director (South East) (2) highlighted the importance of engaging a diverse population. Although their potential use is not yet fully clear, technology presents innovative opportunities to engage diverse populations, including VR, AI and Apps, as was set out in the Planning White Paper (Harris, 2020; Wilson *et al*, 2019; MHCLG, 2020A). Perhaps one of the main reasons why the baby-boomer generation supports the Green Belt the most is because there used to be more planning education, as this planner from the TCPA highlighted:

People used to understand planning better. I suppose there were much wider conversations about it 30–50 years ago. More films, bits of civil society debate, education. You won't find any planning in the curriculum. **So why**

**shouldn't somebody who has grown up with the Green Belt feel out-
raged? How would we expect them to understand what on earth is going
on?…consultation is appalling in this country. It is just dreadful**[14].

To increase public planning education, a basic but consistent level of education, covering the Green Belt, is required in the National Curriculum, perhaps in
Geography or Citizenship, alongside a GCSE or A-Level in Planning to ensure that
younger people can effectively engage with the system (Kraftl *et al*, 2018). As a
pedagogy, it would be a great way to engage children and young people as learning
about a relevant and current process shaping their local community (Olesen, 2018).
Educating the generations between the baby-boomers and children would be challenging but a public education programme, similar to that after WWII, incorporating the latest technology could be incorporated into a national Green Belt 'debate/
conversation', in turn as part of a national plan or Royal Commission on land use.

Other planners argued that local/national politicians should have training in
planning as a minimum.

Education is a key one because a lot of people may perceive any greenfield
land as Green Belt and would be immediately objecting – 'You can't go
there, it's Green Belt!' It's like – 'Fair enough, good point but, actually, you
have got a problem [a housing shortfall] and this piece of Green Belt isn't
serving the five purposes.'…If people understood that, why we have got a
Green Belt or what it does, there might be a better understanding of why
or why it doesn't or can't be released. So, yes, I think education would be
beneficial.

(Private Sector Young Planner (West Midlands) (2))

However, Retired Private Sector Planner (2) argued that mass participation
would be very difficult to achieve, unrealistic and unreasonable:

If you are a single parent, working 8–9 hours a day, trying to look after your
family to keep a roof over your head…are you seriously going to be interested in spatial planning and how things are done? **The answer is going to
be no – it's ridiculous! You wouldn't even consider that. Real, true mass
participation is incredibly difficult. I am sorry to be such a soothsayer of
doom but nevertheless it is what I believe!**.

The validity of this argument is acknowledged, which is partly why embed-
ding planning education in the national curriculum is so important. Nevertheless,
in the absence of leaving planning to the 'experts' with the attendant problems of
democratic accountability (Parker and Street, 2015), it is difficult to see how it can
become more representative of society without wider public engagement whilst
planning deeply affect people's lives (Fainstein, 2014). Additionally, through
Apps such as on whether people 'like' planning applications and which allow 3D
visualisation, wider engagement that is more interactive and less time-consuming

than traditional methods is potentially possible (Wilson *et al*, 2019; MHCLG, 2020A). Likewise, the *Land For The Many* Report (Monbiot *et al*, 2019) and *Raynsford Review* (2018B) made similar arguments regarding having 'Citizens Juries' for planning decisions including a wide range of people and other potential options include a Royal Commission, citizen assemblies, like the UK Climate Assembly, or people's panels as took place in Dublin to decide the location of a very controversial new incinerator (Morphet, 2011). Indeed, education and engagement are particularly important for the Green Belt, which remains widely misunderstood yet has 'capacity' to more effectively engage people in planning (Bradley, 2019A, p. 181).

A National Green Belt Debate

The Brexit Referendum highlighted the importance of education when giving the public a 'say' on key matters but a national Green Belt debate would be a more open-ended 'conversation' on the Green Belt's overall purpose and broad geographical extent rather than a 'yes/no' question (Harris, 2020). It would aim to steer the broader, national debate away from its currently polarised, adversarial nature and, at the local level, from the often narrow focus on specific sites. It would allow the public to think more imaginatively and feel passionate about planning issues with a more positive, abstract, and fundamental debate on broad development patterns in England. There are examples of this in other areas of public policy such as the Citizens Assembly in Ireland over key constitutional questions. The debate could positively incorporate the public's experiential, 'community' knowledge into planning (Bradley, 2018, p. 24), especially as vision is so important in how one views the Green Belt (see Chapter 8). The debate would consider the costs and benefits of the policy allowing people to see its implications and other policies, such as new towns, available as its 'decoupling' from regional policy and new towns has arguably resulted in many people seeing it as a 'shibboleth' on its own and not appreciating its impacts (Mace, 2018, p. 3). For example, if the Metropolitan Green Belt is to remain, there needs to be significant (further) densification of inner and outer London Boroughs but the 'voices' of inner/outer city areas are often not as 'loud' as those in the Green Belt (Haughton, 2017, p. 2; Mace, 2017)[15]. For example:

> **It has become a shibboleth of various ideologues on both sides – those who want to keep it come what may and those who just want to get rid of it. There has been very little careful analytical debate amongst the political classes.**
>
> (Retired Private Sector Planner (2))

> In principle...**I am very happy for the Green Belt to remain entirely untouched if there is a very public open debate about the price we pay for having a Green Belt**...if in London we are happy to say we wish to keep 22% of London as Green Belt and the rest of the Green Belt around

London but the price of that is permanently inflated house prices and the need to build very high densities on land available; everyone has a voice, in principle, I have got no problem with that **but we have never had that conversation**. However, Green Belt wouldn't survive a very open discussion along those lines…so, again, having a conversation about costs and benefits of the Green Belt as a whole and **definitely not as a series of atomised local decisions** – that would be my idealised way of arriving at a new Green Belt policy.

(Planning academic (1))

Consequently, the debate would not solely focus on the policy but include the public's views on their preferred pattern of development, the countryside, economic development and transport infrastructure etc. to 'open up' discussion (Parker and Street, 2015, p. 794; Parker *et al*, 2015, p. 519). Some decisions would have to be arrived at regarding the purpose and overall spatial extent of the Green Belt but this debate could allow the public to feel more engaged and included in planning in the spirit of 'deliberative democracy' (Inch, 2015; Mouffe, 2000, p. 745). Indeed, although there are contrasting viewpoints, visions and values on the Green Belt, a recent 'Citizen's Jury' run by London First, a business organisation, involved giving the jury lots of evidence on the Green Belt and housing crisis then allowed them to arrive at decisions (Community Research, 2019, p. 37; Morphet, 2011, p. 112). 10/11 of the panel supported housebuilding on low-value land in the Green Belt, echoing the findings from the quantitative data that most people take a pragmatic view of it (Mace, 2018; Young, 2019).

In a national debate, the 'silent majority', who support the Green Belt in principle but also limited release of land for housing in particular circumstances, may then emerge allowing a more nuanced Green Belt debate to inform a purpose fit for the 21st century. Once the Green Belt's overall purpose and spatial extent have been established through a high-level national plan, strategic plans could explore various spatial blueprints for regional development with the public, i.e. new towns or urban extensions, and establish locations for development and restraint alongside Green Belt boundaries for the long(er)-term (see Chapter 9). As Planning Academic (1) helpfully argued:

If the political will were there in Central Government to review the Green Belt, I think making it a strategic review has the benefit that you can secure people's trust against reasonable fears that once you start you are not going to stop. You can say – 'Well, this is a one-off review, generationally a one-off review' – then hopefully that will allay some fears.

Conclusions

This chapter has explored why people oppose housebuilding generally and support the Green Belt in particular through critically exploring place, property and principle arguments. It has concluded that, popular planning principles intertwined with place attachment, are probably more important as a motivation than property prices. It has underlined the limitations of the materialistic house price hypothesis, especially neo-Marxian theoretical framework (see Chapter 5), and explored the wider usefulness of the place attachment, environmental psychology and cultural geography literature (Anton and Lawrence, 2014; Davison *et al*, 2016, 2017; Short *et al*, 1987). Whilst this book has focused on a particular Urban Growth Boundary (Green Belt), the findings regarding popular, emotional attachment to place and the countryside have wider relevance to other spatial contexts such as densification, rural development and affordable housing.

The chapter also outlined the lack of trust between campaigners and planners and problems with locally led planning, especially issues with the effective delivery of infrastructure and facilities alongside development, thereby laying the empirical groundwork for the recommendation of rebuilding strategic planning in Chapter 9. It explored the difference between professional and everyday campaigners in terms of their spatial perspective of the Green Belt. It concluded with recommendations based upon the views of planners and campaigners revolving around more planning education, both for politicians and the public, and a national, public and principled Green Belt debate (Inch, 2015). Whilst focused upon England, these findings have wider implications for planning theory and practice internationally with the shared characteristics of community opposition to housebuilding in other geographical contexts, such as Davison *et al's* (2016) work on affordable housing in Sydney or Kjærås (2024) work on densification in Oslo.

Chapter 8 further explores the power of vision and imagination in planning through Lukes (2004) *three dimensions of power*. Nonetheless, the Green Belt, as such a renowned planning policy and the strongest protection against development, is arguably the ultimate 'principle' campaigners lay claim to although support for it may be driven by the deeper, intertwined popular desire to protect the countryside. Alongside place attachment, these planning principles bring out the 'normative' aspects of support for the policy, especially as it commands wider public support than those living next to/in it (Mace, 2018, p. 13). Conversely, the chapter has underlined the importance of place attachment through arguing that 'rationalistic', concerns about the concrete, material effects of development spatially coalesce with normative concerns about a place's character changing in character. At both a principle and place level, it has stressed the importance of the fear of change to campaigners.

These findings, therefore, have more resonance with the broader cultural geography and place attachment literature and concept of 'Homo Democraticus' than that on the house price hypothesis premised upon 'Homo Economicus' (Devine-Wright, 2009; Matthews *et al*, 2015, p. 28). Indeed, this suggests that public deliberation is probably more effective in leading people to reconsider solutions to the housing crisis than direct economic incentives so the politics of planning is explored in Chapter 8 (Inch *et al*, 2020; Sturzaker, 2011).

Notes

1 Surveys on why people oppose development, like CPRE's (2015) survey, are explored later in this chapter.

2 She continued: '*Everyone is a civil engineer because they drive a car and will come in and say: 'Well, this is already a congested area and the road system is going to break'. And they will not proffer any evidence, they will just apply their opinion and say, 'Well, I have to sit at that junction for more than 10 seconds'... they won't look at the fact that the* [Conservative] *Government they voted in cut public spending on the NHS to the bone'.*

3 It would need to be tested in the future to see whether knowledge of the Green Belt 'naturally' increases with age or if it is more *pronounced* among the baby-boomers who had more public planning education.

4 Support varies among superficially similar surveys so it difficult to exactly quantify and elucidate motivations.

5 Unfortunately, data available through the Ministry of Housing/Department for Levelling Up, finished in 2018.

6 An imperfect proxy for this study as not, including factors like attachment to the countryside or Green Belt.

7 When asked whether financial incentives would make one more supportive of development, 66% said there would be no change and 16.9% said that they would be *more likely to oppose* (MHCLG (2019, p. 16)).

8 Fifty-five percent of respondents said that new housing built is *more expensive than existing house prices with only 14% saying that they were cheaper* (MHCLG, 2019, p. 40).

9 Statistics in this section all come from the Social Attitudes Survey 2013 (ScotCen Social Research, 2015).

10 Or 2.2 objections per application, 893 objections daily and 37 objections hourly (Morris, 2019, p. 1).

11 Getting a representative sample of everyday campaigners is challenging as campaigns are often a temporary response to local Green Belt 'threats' whereas professional campaigners are the easiest to interview.

12 7/8 of campaigners interviewed were aged over 60. Arguments about the representativeness of councillors are beyond this book's scope but for critical discussion see: Landmark Chambers (2020).

13 A comprehensive Review of the planning system undertaken by the TCPA (Raynsford Review, 2018A, 2018B).

14 The author visited the International Garden Cities Exhibition (Letchworth) and saw several public information videos on planning and new towns thus confirming the point about the 'baby boomer' generation.

15 See submissions to the London Plan from Bromley (2019) (an outer Borough) and the Telegraph Hill Society (2018) (a conservation charity in Lewisham (an inner Borough)). Telegraph Hill highlighted that the Green Belt is putting huge development pressure on the inner Boroughs, especially its greenspaces, whilst Bromley argued that the distinction between inner and outer London is becoming 'meaningless' due to development pressure.

8   Power, politics and planning

Introduction

This chapter explores the political nature of the Green Belt and planning systems more broadly by synthesising the preceding empirical chapters and drawing out the wider theoretical implications. It examines campaigner's methods of opposing development in the Green Belt and housebuilding generally as elucidated by planners/planning stakeholders and drawing on and critically evaluating the theoretical lens of Lukes' (2004) *Three Dimensions of Power* (introduced in Chapter 5).

The chapter aims to theoretically unsettle the 'given' and 'taken-for-granted' accounts regarding the power of campaigners through examining the contested process of the legitimation and catergorisation of knowledge at a range of spatial scales (Barry *et al*, 2018; Lennon and Fox-Rogers, 2017, p. 366). The chapter highlights that there is an important 'gap' between attempted and effective exercises of power by campaigners, with these exercises being contested at a range of spatial scales, particularly in the process of knowledge categorisation. Crucially, whilst acknowledging temporal fluidity in constellations of power, it highlights the vital multi-scalar dimension as missing from Lukes' three dimensions with actors operating at multiple scales – local, regional and national – trying to exercise and contest power, especially in the production and categorisation of knowledge. There is, therefore, resonance with Allen's (2003, p. 102) work on the importance of space and spatiality to exercises of power with power being 'inherently spatial'. Nonetheless, whilst acknowledging the fluidity and topology of power (see also Allen (2003) and Allen and Cochrane (2011)), the chapter argues that hierarchal scalar power relations are still prescient in public policy.

In policy terms, community involvement is a critical governance and political challenge given the deepening housing crisis, as explored in Chapter 6. Chapter 7 focused on *why* campaigners oppose development; this chapter examines *how* they seek to exercise power, whilst Chapter 9 addresses these themes together by proposing recommendations grounded in strategic planning. This is vital for exploring the *political* feasibility of Green Belt reform (Breheny, 1997; Cherry, 1982, p. 209) with the previous chapters focusing on *practical/economic* feasibility (6) and *social* acceptability (7).

DOI: 10.4324/9781032674315-8

The importance of 'imagination' and 'vision' in shaping how planners and the public view the Green Belt, which is related to Luke's third dimension of power (discourse), is explored in this chapter before returning to address the central research question about *how* different groups exercise power in planning and *which* have the most power (Parker and Street, 2015; Valler and Phelps, 2018; Wargent and Parker, 2018). In relation to the broader, international body of literature on community opposition to housebuilding, it finds that the cultural geography and institutional literature has more utility than the materialistic, neo-Marxian literature when analysing power in planning (Mace, 2018). Indeed, the findings have wider implications with there being shared characteristics in the debate over Green Belt with debates about densification, gentrification, rural development and affordable housing as studies in Sydney (Davison *et al*, 2016, 2017) and Oslo (Kjærås, 2024) have shown.

Tactics: Campaigners Protests and the Politics of Green Belt

The 'Politics' and 'politics' of Planning

Firstly, in focusing on the 'strategies' of campaigners, Sims and Bosetti's (2016, p. 37) framework of 'Politics' (formal politics) and 'politics' (informal politics) is helpful, although spatial interconnectedness is very important with *local* 'Politics'/'politics' leading to the Green Belt being a prescient political issue *nationally*. Most planners accepted that there would inevitably be *local* opposition to development in the Green Belt, with campaigners ('politics') and councillors getting involved ('Politics'). However, they were concerned by the wider political ramifications of blaming *national* politicians ('Politics') and the media/CPRE ('politics') for *escalating local* Green Belt opposition so that it was on the *national* political agenda. The Green Belt being a *national* planning policy, arguably permits the 'jumping' or scaling up of these 'local' campaigns to national politics much more than in countries where the Green Belt is determined at the state or province level, such as the USA and Canada (Cidell, 2005, p. 231; Layzer, 2012).

Firstly, planners highlighted MPs getting involved with Green Belt campaigns as particularly problematic through translating often neighbourhood campaigns into national ones (Brassington, 2019). This appearance of the Green Belt commanding widespread, national popular and political support regularly leads to election campaigns pledges, such as the 'Green Belt is safe under us', alongside hardening political attitudes towards potential reform (Brassington, 2019). Secondly, the regular confusion between Green Belt and greenfield land, the policy's popular emotional appeal and campaigns often being well-supported by certain groups and politicians clearly affect and embolden the tactics of campaigners. For example:

> Because Green Belt is so politically charged, it does energise people more
> in Green Belt areas to oppose and then they are quite ably assisted by their
> local politicians who will also say – 'Yeah, Green Belt is sacrosanct!' Their
> local communities feel – 'Well, if it is sacrosanct, then we can oppose

development and they shouldn't build there!' Then, CPRE comes in with what, at times, are some disingenuous statistics, which we regularly challenge because you can always use statistics in a way which kind of supports your case.

(Private Sector Planning Director (West Midlands) (2)).

Alongside the interaction between 'Politics' and 'politics' locally, campaigners are increasingly resorting to legal action against councils releasing land from the Green Belt, such as South Oxfordshire and the failed litigation regarding Guildford's Local Plan (Lowe, 2020; Young, 2019). As planner from housebuilder (2) argued, this requires significant financial resources and is often concentrated in what she called the 'barrister zone':

The difference when you are in a Green Belt authority (is (litigation)). That obviously gets thrown at you...**more wealthy people tend to live in the Green Belt who have got more monetary and professional resources available to them...ammunition that perhaps isn't available to the 'normal' public. So, when we are looking at Green Belt in certain parts of the country particularly, you will say, 'Well, look, this is a barrister zone'!** So, the protection of views – nobody is entitled to a view – but everyone thinks they are and so **will use things like Green Belt as a tool to hit you with.**

Alongside the social capital that campaigners often possess (Wills, 2016) and the growing professionalisation of campaigning as Roy (2011, p. 591) argued in relation to women's movements in India, campaigners also use more traditional tactics of appealing to people's emotions and employing the 'politics of affect' surrounding popular attachment to the countryside (Thrift, 2004, p. 59). This is often intertwined with support for the policy. For example:

I would take [the Green Belt] out of the local authorities purview and back to national or a regional government because it has become a political football...that local authorities and elected councillors have used to remain in power...residents conflict Green Belt with greenfield and therefore one thing I would do...is change the word 'Green Belt' to 'development belt'. Because, then, the perception and use of the word 'green' would fall away from...if you do that, when it comes to councillors changing and removing land from the Green Belt for housing, then you would not come up against as hard as an opposition as you have now.

(Private Sector Young Planner (West Midlands) (3))

This quote explores the similar dynamics of opposition to both the local plan process and planning applications which could be summarised as 'DAD', 'Decide Announce Defend', and this forms the basis of Table 8.1 (Rydin, 2011, p. 95; Sturzaker, 2011, p. 559, 560).

Table 8.1 Regular sequence of local opposition to proposals to release land from the Green Belt

1. A proposal to release land from the Green Belt for housing is announced in a local plan review or speculative planning application.
2. It is then widely publicised by local campaigners or the media.
3. An opposition group organising petitions, protest marches and placards is formed and publicised through the local press or social media.
4. A local councillor or MP joins in support of the opposition group.
5. This puts pressure on the LPA/developer and, if the proposal is withdrawn, the Group typically disbands claiming 'victory' (unless there is an appeal). If it is accepted, opposition often continues through the Examination, appeal or application process.

Source: Author's own based on interviews.

Tactics surrounding opposition to Green Belt and greenfield development are similar in dynamics but campaigners are perhaps bolder and more passionate about opposing developments in the Green Belt because, firstly, they have a (more) legitimate planning 'reason' and, secondly, can probably draw more support because of the name's emotional appeal. However, other planners, such as Private Sector Planning Director (South East) (1), recognised that this political process is not only inevitable but a legitimate aspect of planning:

There is a lot of politics at the national level. I know there maybe (some) at the local level as well. It is inevitable really because planning is about people and policy. It is about places...And, really, whilst one could be critical of the role of politics within planning, I still think that it is a fundamental part of the planning system because it can bring into play checks and balances. And it is only right that we, as humans, have a desire to protect where we live or agendas that we wish to put forward. Where you have politicians, whether it be Westminster MPs or local/town/parish councillors... they often have a view as to what their sort of local area is right for them and, by and large, the politicians try and represent that view. **I don't always agree with them but that is not the point. The point is that people need to feel they have got representation within the planning system and they are able to express their own views directly through responses to plan-making consultation and applications.**

This last quote is poignant because there was widespread agreement about planning's political nature although differences emerge in planners' attitudes towards it. The close nexus between the local and national and distinctiveness of the policy means that *local* Green Belt campaigns can escalate quickly into *national* ones whilst the Secretary of State and Housing Ministers have significant scope to intervene in local politics. Indeed, with the Planning Inspectorate taking a growing strategic role regarding housing numbers and the Green Belt, it has been the target of increasing political attacks as Boddy and Hickman (2018, 2020) argue although

Young (2019, 2020) highlighted that Ministers have been dissuaded from intervening in lots of appeals/local plans by civil servants in MHCLG (Landmark Chambers, 2020). This underlines the importance of space in researching Green Belt and regional growth management policies more broadly as a crucial interface where local, regional and national politics and protests intersect.

Lukes' Three Dimensions of Power

Having established the empirical basis for this chapter, it now focuses on knowledge and evidence. This is central to many parts and processes of planning systems, especially as the basis for decision-making with knowledge being constructed and classified by planning systems as 'legitimate' – and therefore considered in the decision-making process – or discredited (Bradley, 2019A; Rydin, 2020). The chapter draws upon Lukes' (2004) *three dimensions of power* to conceptually analyse a range of modalities of power in the classification of knowledge. Law (2004, p. 794) has summarised the modalities of power which Lukes' identified – 'authority, domination, seduction, manipulation, coercion and inducement' – but the framework is primarily used in this chapter as providing a useful framework through which to uncover the more overt, coercive forms of power alongside more discoursal, subtle exercises of power (Plaw, 2007; Sturzaker and Shucksmith, 2011). Additionally, the focus of Lukes' (2004, p. 19) framework on the decision-making process and decision-makers, especially 'decisions on issues over which there is an observable conflict of (subjective) interests', is useful given that contested and controversial decision-making in a time-limited context is at the heart of planning systems, in this case in relation to the Green Belt (Rydin, 2020; Sturzaker, 2010).

It is acknowledged that Lukes' framework has been widely critiqued in the literature including the concept of people being ignorant of their 'real interests' in acquiescing with those exercising power (Plaw, 2007, p. 489), for not properly grappling with or defining what power 'is' (Dowding, 2006; Robinson, 2006) and lacking the attribution of responsibility to those exercising power (Morriss, 2002). Whilst this chapter reflects on the crucial, philosophical question of what power 'is', Lukes' *three dimensions* is primarily used for evaluating how power is *exercised* as defining power as the ability 'to force someone to do or not to do something' (Rydin, 2020, p. 214). Lukes' (2004) has also updated his framework to acknowledge more Foucauldian perspectives on power, including viewing power as more dispositional and a 'capacity' (Morriss, 2002, p. 29, 34). Haugaard (2021, p. 153) has recently added a fourth dimension relating to the 'social construction of subjects in response to context'.

The chapter now develops the conceptualisation of the power of campaigning and campaign groups in relation to influencing public policy in planning systems at multiple spatial scales by highlighting that there is often a prescient 'gap' between *attempted* and *effective* exercises of power. These exercises are contested at a range of spatial scales, particularly in the process of knowledge categorisation. Indeed, the classification of knowledge and evaluation of what is 'legitimate' results in the 'production of truth' or 'regime of truth' (Foucault, 1991, p. 79) and categorisation of 'those concepts deemed false from those that are considered true' as 'there is no power relation without the correlative constitution of a field of knowledge' (Lennon and Fox-Rogers, 2017, p. 367). This process of categorisation takes place in many areas of public policy, such as healthcare, social care and policing (for example, Brownill and Inch, 2019; Hurst and Mickan, 2017; Russell *et al*, 2020). Nevertheless, it is particularly important in planning systems which often directly involve the public in decision-making but where 'evidence' and the 'field of knowledge', including national and local policy, is a vital consideration in making time-limited decisions (Dobson and Parker, 2024; Foucault, 1991, p. 79; Inch, 2015).

More broadly, scholars have identified the prescient aspects of this 'field of knowledge' such as the importance of language as the sphere in which 'relations of power are actually exercised and enacted' (Fairclough, 2001, p. 1004). The process of knowledge construction and categorisation is therefore at the heart of planning systems as well as debates about power in planning. Nonetheless, although the Green Belt's political nature has been firmly established in this book alongside that of planning systems generally in other studies (Cherry, 1982; Lord and Tewdwr-Jones, 2018), evaluating *how* effectively power is exercised and *which* groups have the most power was more challenging than expected in the neo-Marxian theoretical frame in Chapter 5. Nevertheless, as Lukes (2004, p. 150) highlighted, evidence and case studies are important in relation to empirically elucidating the *three dimensions*.

First and Second Dimensions: Who Makes Decisions and the Decision and Non-Decision-Making Process

Reflecting Law's (2004, p. 795) point that power is 'complex, multivalent, often conflicting and transversal. Indeed, surprisingly little can be said about its operation in general. Empirical study is required', this chapter examines the specific modalities of power whereby campaigners attempt to exercise power in terms of knowledge production and the categorisation and contestation of these exercises. Reflecting some of the critiques of Lukes' and as the first and second dimensions are arguably closely related (Robinson, 2006), this section explores the decision-making process together in relation to key actors and actants. Throughout, it underlines the circumscription and modulation that there is on the power of campaigners by pro-development actors/actants at both the national and local scale. Indeed, Lukes' three dimensions is useful in uncovering the sometimes 'hidden' way that power is exercised whilst the chapter also recognises that power is always 'inherently spatial' (Allen, 2003, p. 102). Whilst it is acknowledged that all accounts of power are 'partial and limited' (Lukes, 2004, p. 150) and that the data was collected in an English context, arguably there is wider applicability to community opposition in other spatial contexts, such as gentrification, densification, rural development and affordable housing.

Lukes three dimensions are now examined through thematic sections, which explore the theoretical implications and then empirically ground them.

Decision-Making Process I: Attempts to Exercise Power by Campaigners and the Political Process

Although the chapter underlines the categorisation of knowledge and resulting circumscription on the power of campaigners in planning systems, this is not to argue that campaigners do not *seek* or *attempt* to exercise power to influence planning outcomes against new housebuilding (Rydin, 2020). The first key arena of contestation is the political system as, to varying degrees in different planning systems, locally elected politicians are the key decision-makers. In the USA, neighbourhoods can vote for restrictive zonal ordinances with specific densities of housing, thereby excluding lower-income groups (Dehring *et al*, 2008, p. 156; Fischel, 2001A, 2001B). Likewise, in an Australian context, Davison *et al* (2016) have explored how politicians can exploit community opposition to affordable housing as part of campaigning strategies and for political advantage. In England, although the public does not *directly* vote on planning decisions at the local level, Green Belt campaigners still regularly engage directly with the consultation process regarding local plans and planning applications and ultimately, councillors are elected by the local population (Tait and Campbell, 2000).

Knowledge production is crucial at the public consultation stage as this affords the opportunity for campaigners to convince planning officers and local councillors of the validity of their cause and that their concerns are based on 'legitimate' and 'rational' planning reasons[1]. Indeed, as argued later on in the chapter, the increasingly technical and 'calculative' nature of the planning system in England has arguably resulted in campaigners reflexively developing increasingly technical evidence as a way to convince planning officers of the validity of their cause and to try to prove that it rises above sectional self-interest (McAllister, 2017, p. 122).

However, as Roy (2011, p. 591) has argued in relation to women's movements in India, alongside the 'professionalisation' of campaign groups, there is still evidence of the 'passion' related to campaigning causes and making an 'emotional' case regarding the legitimacy of their claims to knowledge. Indeed, this arguably relates to Thrift's (2004, p. 57) arguments around the increased importance of the emotional and the 'politics of affect'. In England, the abolition of statutory regional planning and strategic decision-making regarding the Green Belt since 2010 has arguably increased the political nature of planning decision-making, with *local* councillors now largely responsible for Green Belt release (Goode, 2022B). This has meant that the exercise of power has 'jumped down' to the local scale (Cidell, 2005, p. 231) in terms of the Green Belt gaining more prominence as a local political issue. Nevertheless, in relation to knowledge categorisation and building the case for the legitimacy of their cause, it also has led to the 'jumping up' of campaign groups in relation to knowledge claims, i.e. that a specific Green Belt development represents an emblematic threat to the principle of the policy and to the broader institution of the English countryside (Goode, 2022A).

On the one hand, this demonstrates the importance of spatiality in campaigning in that, whilst the Green Belt is a national policy, issues get *refracted* or intensified by local circumstances, underlining the importance of spatial variation and place specificity in the exercise of power (Allen, 2003; Amati, 2007). On the other hand, whilst somewhat reflecting the specificities of the localised nature of the English planning system, this demonstrates the importance of spatial interconnectedness, especially the interconnectedness of the local and national in campaigning, as national politics and policy shape local campaigns. However, campaigners seek to 'scale up' and generalise their claims to knowledge in order to appear legitimate (Cahill, 2004; Roy, 2011). Indeed, whilst there has been debate about the relevance and validity of 'scale' as a concept in Human Geography (Marston *et al*, 2005), it arguably continues to be of prescience in the realm of public policy, especially planning systems, which are characterised by subsidiarity and hierarchal (spatial) policy scales in decision-making (Natarajan, 2019).

Empirical Dynamics I: Locally Led Planning

As argued above, the abolition of Regional Spatial Strategies increased the political nature of planning decision-making, with *local* councillors largely being responsible for Green Belt release (see Chapter 9) led to campaigners *attempting* to exercise power both directly through campaigns and indirectly through elections (Goode, 2022A). This mechanism is elucidated by the Private Sector Planning Director (South East) (2), who also referred to the 2019 Local Elections:

> Strategic planning is something the system needs…inevitably, when you are making big controversial decisions, if as members you are going back to the politics…in that district, you are going to be affected to a greater or lesser extent by that decision and lobbied by people who live in the district, it is very difficult to make difficult decisions.

There is pressure on LPAs by the national Government to produce up-to-date local plans, so several Conservative councils adopted plans with significant Green Belt release in order to be found 'sound' by the Inspectorate (Wilding, 2018). Although these were arguably politically brave decisions, a significant political backlash resulted in the Conservatives losing control of councils in places such as Tandridge, South Oxfordshire and Guildford, primarily due to opposition to local plans by Resident Groups and the Liberal Democrats (Branson, 2019). This clearly demonstrates the effectiveness of the 'Politics' and 'politics' of campaigners in exercising power through seeking to change *who* makes decisions via electoral pressure and campaigning.

A planner in Focus Group I also highlighted the age of councillors making decisions:

> Our decision-making strategies are skewed very badly. The representation of people who make the decisions tends to be around the age of 60 rather than representatives of those who at the time of their life when they are probably most open to change and new ideas (aged 20-35).

Finally, the following quotes explore the increasingly politicised nature of decision-making in the era of localism in the West of England:

> Probably the lowest moment was when the famous Eric Pickles [Communities Secretary] indicated the intention to abolish regional planning...there was research [Tetlow King Planning] looking at authorities that slashed their housing requirement and all of the four West of England authorities were in the top ten. So, I think there has been a desire to protect the Green Belt... You could accuse the authorities of not working together effectively...the biggest input into those strategies was probably what is politically the most acceptable or politically least unacceptable. So, in Bath and North East Somerset, the politically acceptable has resulted in a Green Belt allocation, adjacent to Bristol, not near Bath...In North Somerset, the politically acceptable words were 'leapfrogging' the Green Belt because the Administration some time ago had said 'No Green Belt Review over our dead bodies! We are not meeting the needs of Bristol – we are North Somerset!'...So, **in the increasingly politicised planning system, it seems to be the easiest political thing to latch on to was the most well-known planning designation**...Any planning system can stop new homes if there is a political will to do that!
>
>   (Private Sector Planning Director (Bristol) (1))

> **It is almost political suicide for any local authority to revisit Green Belt unless they are dictated to do it**...North Somerset didn't join the West of England Combined Authority for that very reason – they were struggling with the challenges on their release of Green Belt.
>
>   (National Private Sector Planning Director (2))

These quotes encapsulate the politicised nature of decision-*makers* and the decision-making *processes* regarding the Green Belt. Whilst the policy is a *general* political issue, it appears that it is *refracted* or intensified by place specific circumstances, reflecting the importance of geographical variation in how LPAs manage 'their' Green Belt. This includes their institutional culture, as evident in the West Midlands and West of England[2] (Amati and Yokohari, 2006).

Decision-Making Process II: Contestation of the Exercise of the Power of Campaigners by Policy and Legal Processes

Most planning systems have a basis in law with both national policy and legal processes. This is particularly the case in zonal planning systems where ordinances set the parameters of land use and development, but discretionary planning systems, such as the English one, also have legal tests and the right to appeal decisions (Fischel, 2001A, 2001B; Inch *et al*, 2020; Tait and Campbell, 2000). Whilst varying in the degree of detail and prescription, national planning policy is a prescient way that the legitimacy of campaigner's knowledge is both 'scripted' and 'categorised', including what is considered as a 'material consideration' in terms of objection and therefore counted as legitimate or 'warranted knowledge' (Rydin, 2020, p. 214). Indeed, in many ways, national planning policy sets the parameters or 'rules' by which campaigners need to abide. For example, in England, there has been a 'presumption in favour of sustainable development' since 2012 in national planning policy – the National Planning Policy Framework (Inch *et al*, 2020, p. 724).

This is reinforced in many countries by legal processes and judicial structures which aim to uphold and enforce national planning policy. In England, local plans are assessed in the degree to which they confirm to national policy by the quasi-judicial Planning Inspectorate (McAllister, 2017, p. 122). Moreover, if a developer feels that a local authority has unfairly refused planning permission and not acted in accordance with policy, there is the opportunity to appeal to the Planning Inspectorate with the possibility of local rejections being overturned. Indeed, the decision-making process is laced with the threat of appeals and circumscribed by national policy (Goode, 2022A; Tait and Campbell, 2000). Unlike other countries, such as the Republic of Ireland, there is no community third-party right of appeal against the granting of planning permission.

One result of this is the professionalisation of campaigning and increasingly technical evidence being used at planning committees – where local councillors vote on local plans and planning applications – as well as at planning inquires where local plan examinations and appeals are determined by the Inspectorate. In order to ensure that their knowledge gets categorised as legitimate, campaigners are developing their own housing/population growth studies and challenging 'official' reports such as the GL Hearn Growth Study in the West Midlands.

Campaigners and councillors, therefore often follow a line of 'agonist' resistance through sometimes working against the 'system' yet still aiming to secure concessions from it with local authorities and elected members often 'sandwiched' between community opposition and housing targets (Lennon, 2017, p. 154; Inch, 2015, p. 418). Indeed, whilst Green Belt campaigns *attempt* to exercise power through seeking to influence politicians, there is also the 'shutting down' or 'silencing' of these dissenting voices, with local authorities blaming national Government when justifying very unpopular Green Belt release (Parker and Street, 2015, p. 794). More broadly, this again underscores the importance of hierarchal scale in planning systems and, through examining the 'actual workings of power' or 'practices of power', illustrates Allen's (2011, p. 291) argument about how actors can be geographically distant yet exercise power in an immanent and 'real time' way. Indeed, arguably the abolition of regional planning in England has led to increased central Government intervention in order to circumscribe the power of campaigners and discipline councils which deviate from national policy, such as Sevenoaks and South Oxfordshire (Inch *et al*, 2020). Additionally, in terms of campaigning strategy, it demonstrates how campaign groups can pragmatically apportion blame and responsibility to the national Government as part of their 'performativity of practices' (Rydin, 2020, p. 217).

Empirical Dynamics II: National Planning Policy, Planning Inspectorate and 'Planning by Appeal'

There are important limitations or counterbalances on how effectively politics and campaigner pressure can determine planning outcomes, especially with growing popular calls to 'solve' the housing crisis exemplified in the YIMBY movement (Davis and Huennekens, 2022; Myers, 2017).

Firstly, where an LPA lacks an up-to-date local plan, does not have a 5 Year Housing Land Supply (5YHLS), fails the Housing Delivery Test, or a planning committee refuses an application against officer advice, there is the possibility of developers 'appealing' against the decision, which is often successful on greenfield and, sometimes, Green Belt land (Young, 2019, 2020). A planner from a West Midlands land promoter, previously an LPA planner, explained:

> I have sat in many planning committees with placards being waved at the back of the room and you see how that impacts the decision being taken and, you know, the amount of times I have seen recommendations for sites being overturned by the Committee and then granted on appeal – **it is part of a political game!** And, ultimately, whilst the planning system retains a largely political element in terms of that decision-making, it is very difficult to see how you arrive at a solution where we actually build the amount of houses we need to in this country because you are putting decisions into the hands of, yes, democratically elected people, but that's at the heart of the challenge... We had a recent refusal in an LPA where the emerging local plan says it needs 40 homes in this settlement.

Similar points were made by a planner in Focus Group II:

> You wonder if this lack of governance at the regional level is going to lead to more speculative applications and appeals coming forward in the Birmingham area from developers who are getting fed-up with waiting. They have got options on land or maybe even own land in the Green Belt but they are willing to take a punt on it because, in this way Birmingham and the surrounding authorities aren't really going to be able to demonstrate a 5YHLS.

However, Private Sector Planning Director (South East) (2) offered some balanced reflections:

> We know from appeal decisions and High Court judgements as well, that housing land supply itself is not a justification for development in the Green Belt. But Green Belt release for development is justified if it comes through the local plan process... by devolving that responsibility, firstly, to local politicians and their officers. Then, obviously, it has got to go through a local plan examination where you have an Inspector who has to go through the same balancing exercise... Ministers generally are reluctant to get involved

in a development plan system. They get involved in appeals all the time...
[however] 'planning by appeal' – there maybe some promoters who would
look forward to that but most planners wouldn't.

As this quote and Private Sector Planning Director (South East) (1) highlighted,
the Secretary of State can overturn appeals (Edgar, 2015). Nonetheless, the appeal
mechanism and 5YHLS exert significant pressure on councillors *against* community opposition and is an important way that the development industry exercises
power. Indeed, perhaps power in planning could be better conceptualised as *fluid
and contested* in a temporal and geographical sense in that housebuilders *and*
homeowners attempt to exercise power, but it is difficult to define *who* exercises
power most effectively, with the planning system being where these conflicts work
out and are resolved.

Decision-Making Process III: The Categorisation of Knowledge by Planners

Community engagement has a statutory underpinning in many planning systems
and forms an important part of the training and education of planners (Brownill and
Inch, 2019). However, the degree to which planners regard campaigners' objections
as genuine and knowledge as legitimate is another important part of the process of
categorisation in decision-making. As Chapter 7 highlighted, in planning systems,
in many ways, there is a conflict or tension between 'planning' knowledge, which
is associated with rationality, the expertise of planners as professionals and viewed
as technical, objective or even 'scientific' and the more emotional, experiential and
tacit 'community' knowledge associated with campaigners (Bradley, 2018, p. 24).
Moreover, this is related to the 'public deficit explanation' or planners can resort to
the 'public-deficit explanation' of the public as laypeople without an understanding
of how planning systems operate (Welsh and Wynne, 2013, p. 552). Additionally,
as Chapter 5 demonstrated, there is the widespread characterisation of campaigners, particularly against new housing development, as selfish, irrational 'NIMBYs'
(Dehring *et al*, 2008, p. 156; DeVerteuil, 2013, p. 599).

   Whilst in many planning systems, there is a statutory obligation to take account
of objections, the internal categorisation process of planners and the use of 'professional judgment' can lead to the ignoring of these dissenting voices (Rydin, 2020,
p. 225). Indeed, planners often view the 'performance' of objecting as 'merely
a formality' and a threat to the 'objective' basis of decision-making and broader
'dispositional rationality of planning' (Rydin, 2020, p. 225). Clearly, this process of
categorisation varies between professional planners, but arguably, there are shared
viewpoints or metanarratives regarding the characterisation of planners as 'experts'
juxtaposed to the public as 'ignorant' and 'irrational' (Welsh and Wynne, 2013,
p. 552).

   The broader literature has arguably not paid enough attention to the specific
modalities of power in planning systems and, as unsettling 'taken-for-granted' accounts (Barry *et al*, 2018; Lennon and Fox-Rogers, 2017, p. 366), Lukes' *three*

dimensions is useful in highlighting the sometimes more hidden ways that power operates in relation to decision-makers and the decision-making process (Sturzaker, 2010; Sturzaker and Shucksmith, 2011). However, perhaps understandably as coming from a Political Science background, arguably Lukes overlooks the importance of space in the exercise of power, especially hierarchal spatial scales, which this chapter has underlined whilst acknowledging the contested, complicated and 'messy' nature of power in line with Allen's broader arguments. Whilst beyond its scope, it acknowledges the *fluidity of power* in a temporal sense in that the respective power of housebuilders *and* campaigners can vary over time depending on national policy and local political control, etc. (Short *et al*, 1987).

Empirical Dynamics III: The Plan-Led System and the Consultation Process: Agonist Planning

An example of the cynicism of planners can be seen in the following quote which, in many ways, encapsulates the views of planners

> The high level of, shall we say, engagement with the local plan process, I think has been brought on by the fact that, broadly speaking, it is one of the wealthiest areas of the country and very much on that Green Belt fringe… perhaps not unsurprisingly, areas that are most affected by…Green Belt release, have been the most vociferous and early represented individuals in the consultation process, including attending the examination etc. and in person themselves!…there is an affected group of villagers, so to speak, that are particularly concerned because their lovely twee house in the countryside is about to be destroyed by housing nearby…The Green Belt allows people to kind of attach…their negative attitude towards something concrete…'We don't want anything around us because we like our nice countryside walks and view from the back of my house' which is something that is a legitimate part of national discourse.

Whilst this reflects the intensity of place attachment and shows how the consultation process of decision-making is dominated by particular groups, i.e. housebuilders and campaigners, the quote also demonstrates how these attempts to exercise power are often viewed sceptically by planners, undermining how far campaigners can influence material planning outcomes. Moreover, even councillors making decisions on planning committees arguably have limited scope to be *overtly* political, with the decision-making process being laced with the threat of appeals and circumscribed by national policy (Goode, 2022A; Tait and Campbell, 2000).

These direct campaigns alongside broader electoral campaigns against development can be fruitfully conceptualised as 'agonist' resistance, with councillors and campaigners sometimes working against the 'system' but still aiming to secure concessions from it (see Chapter 8) (Parker and Street, 2015, p. 794; Vigar *et al*, 2017, p. 426). LPAs are often 'sandwiched' between community opposition and housing targets. Indeed, whilst Green Belt campaigns *attempt* to exercise power through seeking to influence politicians, there is also the agonistic 'shutting down', 'silencing' or circumscribing of these dissenting voices with LPAs blaming national Government when justifying very unpopular Green Belt release, such as in Solihull, Tandridge and Guildford (Goode, 2022A; Parker *et al*, 2015, p. 519)[3].

LPAs which have tried to reduce Green Belt release in local plans following the voting out of local Conservatives, like South Oxfordshire, or where Conservative LPAs have tried to limit Green Belt release, such as Sevenoaks, difficulties have arisen with the Inspectorate or they have been forced into adopting plans by national Government (like South Oxfordshire) (Donnelly, 2019; Kahn, 2020)[4]. This book has demonstrated the lack of power and disenchantment campaigners often say they feel with clear circumscription on *how far* campaigners and politicians can *effectively* exercise power with the Government heavily, and increasingly, involved in the decision-making process (Goode, 2022A; Landmark Chambers, 2020). Conversely, since 2010, the blame for controversial decisions on the Green Belt has been deflected, dissipated and redirected by national to local Government with the policy arguably being increasingly used by politicians locally and nationally as a form of statecraft and a tool to gain, maintain and retain power (Mace, 2018, p. 5).

The aim of this chapter is not to assess the validity of comments by campaigners about their powerlessness but, in terms of the modalities of power, it demonstrate how in the decision-making process in planning systems, there is often circumscription on *how far* campaigners can *effectively* exercise power. This includes the 'shutting down' or modulation by 'absent others' (Parker and Street, 2015, p. 794; Rydin, 2020, p. 228).

Empirical Dynamics IV: The Local and National: National Policy and Local Autonomy

The local plan examination process and the pressure and representation by housebuilders throughout also place important limitations on the *effective power* of campaigners. Professional campaigners regularly argued that the system was skewed in favour of developers:

> Since the NPPF...we are getting development completely throwing out the Green Belt policies. An example of that is Solihull Council in their Local Plan Review – they started suggesting new housing developments before they had done the Green Belt analysis. Now that was completely wrong.
>
> (Planner (3), CPRE West Midlands)

The close nexus between the national and local in the exercise of power by housebuilders was highlighted. Indeed, the housebuilding lobby is another significant counterbalance against community opposition, with planning as the contested arena where this conflict is played out (Short *et al*, 1987). Some planners argued that the system could more effectively counterbalance campaigner opposition with Private Sector Planning Director (Bristol) (1) highlighting:

> Through the independent role of the Planning Inspectorate,…the soft and hard power of the Secretary of State in MHCLG…you could…see a way to recognise the housing crisis as they do now, but also recognise that the Green Belt is a constraint on building new homes – not just the amount of new homes but new homes where they need to be dotted around our economic powerhouses.

Nonetheless, whilst planning's political nature is evident and frustrated planners, especially *national* politics, politics in planning is probably inevitable and perhaps to some extent desirable as a way or outlet to maintain public confidence in the system as this insightful remark by Private Sector Planning Director (South East) (1) shows (Wannop and Cherry, 1994; Goode, 2022A):

> Politics is very important…There are some who would say that politics just intervenes and doesn't necessarily add any value to the process, especially if you have just finished a planning committee at 10 o'clock at night. You are recommended for approval but councillors have voted it down…you can feel quite sore. But I think when you rise above that and look more generally – there is a lot of value that sort of representation can add to the planning process and it is right at the national level there is a [Green Belt] policy as well.

Some planners suggested abolishing planning committees, especially for planning applications with more delegated powers for planning officers (a similar proposal to what was in the Planning *White Paper*). However, they are vitally important for giving planning democratic accountability and the public confidence that applications are properly scrutinised (Goode, 2022A; MHCLG, 2020A, p. 37; Tait and Campbell, 2000; Yarwood, 2002). Nonetheless, whilst the presence of politics in planning is recognised and inevitable, the *amount* of politics could arguably be reduced, so Chapter 9 considers how the system could become *less politicised* through strategic planning.

The Third Dimension

DISCOURSE, METANARRATIVES AND EVIDENCE

Lukes' third dimension underlines the discourse used by dominant groups so that decisions are accepted by people through 'willing compliance' as the 'natural' order of things' – it is, therefore, the most 'insidious' exercise of power (Lukes, 2004, p. 28; Rydin, 1985, p. 65,, 2020, p. 217). Whilst more subtle and abstract, discourse and securing public acquiescence can be a powerful strategy that both planners and campaigners utilise in attempting to exercise power (Lukes, 2004; Sturzaker and Shucksmith, 2011). More broadly, Thrift (2004, p. 57) has underlined the importance of the emotional and the power of the 'politics of affect'. The chapter has already outlined how, in many ways, the discourse of campaigners in England has become increasingly professionalised through now commissioning their own technical household projection studies etc. as trying to empirically 'disprove' official 'inflated' household figures and therefore have their knowledge categorised as legitimate. Additionally, Rydin (2020, p. 222) has demonstrated how affluent communities are more able to marshal the necessary resources to build such evidence, which is treated as more legitimate in planning systems. Nevertheless, the more traditional, emotional use of discourse as an exercise of power arguably remains important in many ways as campaigners and housebuilders try to develop very powerful affective and, in some ways, binary or competing metanarratives, paradigms or coherent frames of vision in order to persuade the public of the moral validity of their cause (Warren and Clifford, 2005).

This is particularly the case in England with the spatial interconnectedness between the local and national in the planning system with campaigners and politicians building a general narrative around the intrinsic beauty of the English countryside, protected by the Green Belt, which should be guarded as a principle. The corollary of this is that most housebuilding should be in existing cities on brownfield land with an arguably idealised view of urban living. This remains an enduring and very popular metanarrative which is anchored around the affective emotional appeal of the English countryside embedded in the 'common mind and language' (Rydin, 2020; Warren and Clifford, 2005, p. 217). More broadly, the countryside and Green Belt is powerfully presented to the public by campaigners as knowledge which is the 'natural' order of things' (Rydin, 1985, p. 65) and further that this 'field of knowledge' appears as a 'regime of truth' (Foucault, 1977, p. 27).

However, consistent with the 'conceptual binary' outlined above of technical 'planning' knowledge being juxtaposed to emotional 'community' knowledge (Welsh and Wynne, 2013, p. 552), this discoursal attempt at exercising power through discourse is arguably increasingly contested by developers and planners developing the 'rational' argument around more land being needed to 'solve' the housing 'crisis'. In terms of knowledge categorisation, this aims to discredit both the validity of the 'emotional' argument of campaigners and build a convincing alternative metanarrative. This argument does not discount the intrinsic beauty of the countryside and Green Belt but it argues, in rationalistic terms, that much of the countryside remains undeveloped and that only a small amount of land is needed for building to 'solve' this deepening housing 'crisis'. Indeed, increasingly the language of 'housing emergency' is used underscoring the urgency of action according to this metanarrative and this has found political resonance and support, particularly in the Labour Party, with the Prime Minister, Sir Kier Starmer's comments about building on the 'grey belt' (Martin, 2024; Young, 2023, p. 1). Of particular poignancy to this argument is the widespread disparaging of campaigners against development as irrational, selfish 'NIMBYs' largely concerned parochially with their local area, which discredits their knowledge and undermines their discoursal attempts in the exercise of power (Dehring *et al*, 2008, p. 156; DeVerteuil, 2013, p. 599).

Another powerful corollary of this argument is more dystopian visions of the Green Belt causing 'town cramming' and 'rabbit hutch homes' and an alternative vision of more 'sustainable' urban extensions (Mace, 2018, p. 9). However, the 'regional' spatial dimension is also very important here, particularly in post-industrial regions such as the West Midlands, where the scale of urban growth during the Industrial Revolution, alongside the heavy nature of its industry, continues to feature strongly in the popular imagination, especially in terms of the threat of 'encroachment' by the conurbation (Goode, 2022B). This relates to the broader characterisation of the 'urban' or cities being 'dirty' and 'evil' as juxtaposed to the 'purity' and 'pristine' character of the 'rural' or the countryside (Warren and Clifford, 2005, p. 585). However, because the Green Belt is a national policy and there is wider popular concern about the countryside as an institution being 'under threat' by development (Inch *et al*, 2020), campaigners can relate and 'jump up' these 'local' campaigns to broader, powerful metanarratives again highlighting the importance of spatial interconnectedness in the exercise of power (Cidell, 2005, p. 231).

In relation to the power of discourse and convincing metanarratives, the views of campaigners are exemplified by this quote by a national Conservative politician interviewed:

The original purposes of the Green Belt are about beauty and the sense of communities as meaningful places rather than sprawl without an identity – we want to be a country in which our towns, cities and villages have an individual identity, are distinct from each other and there is greenery between them – you can favour that living in the centre of a city.

In terms of the importance of the relationship between the local and national in campaigning, a Save Stourbridge Green Belt campaigner interviewed argued, in a quote that exemplifies campaigner perspectives:

'There is so much opposition to development of green belt around Stourbridge and England in general because people realise that it is valuable resource that could be very easily lost forever…I am particularly concerned that the green belt is preserved to prevent urban sprawl and to protect the integrity of existing towns.

In terms of the 'rational' or 'pragmatic' arguments made by planners and developers, the following quote exemplifies a widely held professional/practitioner viewpoint:

There is probably…a dilution in the purpose of Green Belt policy and, for many local authorities as well as opposition groups, necessarily some misunderstanding or misrepresentation, as I see it, of the real purpose…Originally, it was to restrict urban sprawl.

(Young Planner, National Housebuilder)

This arguably reflects wider debates in society regarding the lack of 'control' people feel that they have over the change in their local communities (Goode, 2022A; Inch *et al*, 2020), perhaps reflected in the Brexit vote. Nonetheless, it appears that the Green Belt is a very significant *battle of ideas or discourse* between campaigners and planners/developers or a clash between the 'head' and 'heart', i.e. campaigners advocating for the countryside ranged against planners/housebuilders arguments for 'solving' the housing 'crisis' in a 'rational' way (Bradley, 2018). Both these groups deploy affective discourse reflecting these very powerful metanarratives as ultimately attempting to persuade public opinion locally and nationally of the moral validity of their cause and that it is in the 'national interest' (Warren and Clifford, 2005, p. 587). Indeed, many broader and profound societal issues, including the importance of the countryside as an institution to national identity and the serious problem of intra-generational inequality in housing (which a range of think-tanks, academics, media commenters and planners/developers view the Green Belt as key to solving), seem to coalesce and find expression in relation to the Green Belt.

In many ways, this therefore demonstrates how the third dimension of power is subtle but perhaps the most powerful way of exercising power demonstrating the utility of Lukes' *three dimensions* (Dowding, 2006; Lukes, 2004; Rydin, 1985). Of course, the tension between development and conservation ideas has arguably *always* existed in many planning systems but post-war planning policy in England aimed to reconcile it through meeting housing growth via new towns, regional policy and then a brownfield-first policy alongside the Green Belt protecting the countryside (Mace, 2018). However, with deepening housing affordability problems and the localism agenda, these competing visions recently have been more vigorously fought out and had to be reconciled at the local level (Inch *et al*, 2020).

Whilst the above particularly relates to an English context, arguably the contested discourse of planners/housebuilders and campaigners and the development of meta-narratives is applicable in a range of geographical contexts where there is conflict over greenfield housing development such as the Groene Hart ('green heart') in the Netherlands, Urban Growth Boundaries in places such as Portland and indeed the failure of Green Belts in South East Asia, Japan and New Zealand (Altes, 2017; Amati, 2007; Boyle and Mohamed, 2007). Likewise, there are shared characteristics with the debate over the 'emotional' case against densification as being out of 'character' of an area and the case for development of maximising the use of urban land and delivering affordable housing as studies in Sydney (Davison *et al*, 2016) and Oslo (Kjærås, 2024) have shown, for example. Moreover, whilst focusing on a different character of debate, there is also resonance with debates about gentrification and the case made against it in terms of social displacement and fundamentally changing the character of areas and the market rationality in favour of it in terms of (it is argued) that there not a viable alternative in the current context (Kjærås, 2024). This underscores the broader importance of the categorisation and contestation of knowledge production in terms of discourse and construction of metanarratives.

The Third Dimension – Empirical Dynamics

The empirical basis of the competing meta-narratives has already been explored so this section presents the empirical material on the professionalisation of discourse by campaigners before examining how planners 'framed' the past.

Discourse and the Strategy of the Professionalisation of Campaigning

Professional Campaigners

Even before planning became increasingly technical and evidenced-based as part of the 'calculative turn' since 2010 (McAllister *et al*, 2016, p. 2363), Amati (2007, pp. 585–591) highlighted that the majority of objectors to Green Belt development were 'experienced' as able to use planning's technical language to assert 'technical kinship' with planners to try to persuade them of the legitimacy of their concerns. However, campaigners are now commissioning their own increasingly technical household projection studies etc. and regularly challenging estimates of much housing can be accommodated on brownfield sites as too 'low' thereby arguing that the need for Green Belt release can be significantly reduced or eliminated[5].

For example, Planner (1), CPRE West Midlands, argued:

[You need to be] careful of how you assess housing needs…we [CPRE] are pretty critical of the GL Hearn Study[6]…**it contains a lot of arbitrary assumptions**…Some of the problems though are about Government policy on assessing housing need rather than being specific to the GL Hearn…CPRE nationally has expressed its views on the NPPF/PPG…[I have pointed out] the absurdity of the Government's position on this[7]. I don't think we should take the claimed scale of housing need as a given, whether nationally or in a particular area. We should be prepared to assess it. What CPRE would like to do is what we used to, work with the local authorities in that process.

Similar critiques were levelled by this Policy and Campaign Advisor, West Midlands, who explored the relationship between housing figures and the Green Belt before criticising the BDP's for its unmet housing 'need' figure of 38,000 and underestimation of potential from 'windfall' sites (Birmingham City Council, 2017, p. 144):

The 2016 ONS [figures]…have changed. People aren't living as long as we expected…but, most importantly, the size of household isn't decreasing as much as we expected…It is to do with when young people leave home and all those kind of issues. The ONS's latest figures dramatically reduce urban need as a result of that change. The Government…have said 'never mind it being the newest evidence' which they have never said before. They have always said it when it goes up and now they say you can't take that evidence on face value…So this whole housing shortfall is the creation of figures that now are pretty well-inflated and, **there maybe a shortfall, but it is not nearly as big as they say and, the trouble with having too much of a shortfall, is you then allow the exceptional circumstances route to be used and release too much land in the Green Belt**...The Public Enquiry is actually out of date.

Unsurprisingly, professional campaigners interviewed did not criticise population growth or immigration *per se,* probably to focus on legitimate 'planning reasons' (Hubbard, 2005; Sturzaker, 2010). In fact, the only mention of migration was in passing about population projections by Retired Strategic Planner (1).

Everyday Campaigners

However, these arguments came up among everyday campaigners, especially the SSGB responses:

You should start with the root cause and pressures for housing, that is Population Growth and a subset of this is the breakdown of society and the family unit putting greater pressure on precious land, be it labelled Green Belt or otherwise.

Social housing...that's what is so desperately needed. Controlled immigration, sensible use of existing stock.

The complicated, highly controversial question of how far migration is responsible for population growth and the feasibility and impact of policies, such as 'capping' population and 'controlling' migration, are beyond this book's scope but it clearly features with some, largely everyday campaigners opposing Green Belt development. This resonates with Sturzaker's (2010, p. 1014) work on snobbery and the 'exclusionary preferences' of the wealthy in rural areas. Although SSGB is only one campaign group, it is perhaps a more widely held position with Private Sector Planning Director (West Midlands) (2) explaining:

Some people want to leave the EU but, when you press them, they don't know why...Sometimes that is the same attitude towards development which is – development is a bad thing...it might be even immigration based. I have heard people opposing new development on the basis (this was in Coventry) – 'There are too many Poles and Romanians in this city already and the only reason we are having to build more houses is to accommodate them'. And I don't know quite how to react to that. Is that just purely racist or equally, turning it on its head, can you understand people's concerns about migration and the impact it has differentially on cities and towns?.

These sentiments are often drawn from and echoed nationally by politicians, with the then Housing Minister, Dominic Raba, stating in 2018 that fewer houses would need to be built in the Green Belt if migration was effectively controlled post-Brexit (Geoghegan, 2018). Immigration and population growth are often easy scapegoats for the housing crisis, but arguably, the *underlying* reason for campaigners opposing development is fear of change, especially the lack of control people feel that they have about their local area changing in character.

Reflections on Campaigners' Strategies and Tactics

The reflexive move of professional campaigners towards technical evidence shows their flexibility and how they constantly evolve to stay 'relevant' in different planning contexts (Amati, 2007; Bradley, 2019B). CPRE has been particularly successful as a campaigning organisation in fighting a 'ceaseless war' against development and in favour of the Green Belt (Amati and Yokohari, 2003, p. 321; Manns, 2014, p. 8). This resonates with the flexibility of the policy itself with justifications for it varying temporally in different economic and political contexts, i.e. urban regeneration being added as an objective in the 1980s (Amati and Yokohari, 2003, p. 321; Mace, 2017; Manns, 2014, p. 8). Another example of campaigners flexibility and alacrity is the increased emphasis recently on climate change, air pollution and the importance of greenspace and the Green Belt following the Coronavirus lockdowns (Goode, 2022B; Harris, 2020; Mell, 2024)[8]. Of many examples, campaigners against the Guildford Local Plan argued that Green Belt development would 'impact on the already poor air quality around Compton' (Curley, 2019, p. 1).

It appears that a lot of time, energy and concentration is spent on debating housing figures rather than focusing on equally important issues, especially strategic development locations and their design, facilities, infrastructure etc. (Young, 2021). Although RSSs were criticised for 'forcing' housing numbers on LPAs, there is a clear case for national Government setting regional targets to remove local wrangling, as was proposed in the Planning White Paper and is proposed in the NPPF (2024), with this being the case for most of the post-war era (Best, 2019; MHCLG, 2020A)[9].

'Framing the Past': Retired Planners Views

Finally, the power of discourse can be seen in the nostalgic view that older, retired planners often took of planning policy and practice, especially of the Green Belt. Of course, there are important lessons to be learned from the past, for example, about regional planning and new towns, but, as Valler and Phelps (2018) argued, the dominant, often powerful, planning ideals about development patterns can also restrict future vision. For example, as a Retired Planner (South West) pointed out regarding the West of England Joint Spatial Strategy:

> Well, the new spatial plan, is exactly the area of Avon…And you get the two documents out and it is just rewritten!

Many of the planners most supportive of the Green Belt and regional planning were retired with planning experience *before* the era of 'growth dependent planning', which began during the 1980s and continues to the present day (Lord and

Tewdwr-Jones, 2018; Rydin, 2013, p. 3). They are also part of the baby-boomer generation among which the Green Belt commands particularly high levels of support (see Chapter 7). Indeed, the Green Belt, as associated so closely with the post-war planning system, almost seemed interwoven with their professional identity. Many of them spent most of their working lives in the public sector and were driven by an ethos of public service (Slade *et al*, 2019).

Many of the middle-aged planners, typically in their 40s and 50s, were particularly critical of the Green Belt. Their professional experience has largely been of growth dependent planning in which they have been acculturated, typically working largely in the private sector, and they had more awareness of the housing crisis, often through its effects on their children (Raco *et al*, 2019). Young planners aged in their 20s and 30s tended to be not so outspoken in criticising the Green Belt, perhaps reflecting the fact that they were earlier in their careers and not so confident of expressing their professional 'voice' (Slade *et al*, 2019, p. 33). Clear generational differences can, therefore, be seen among planners in their direct attitudes towards it, with the potential impact of age, professional identity, sector of work and wider societal attitudes upon these views alongside personal/professional values being a very interesting area of potential future research although beyond this book's scope (Kenny, 2019B). Nevertheless, a range of views were still expressed even among planners of similar ages and there was not uniform, linear progression in support of the Green Belt according to age as this quote by a young planner in Focus Group II illustrates:

> The Green Belt is always seen as sacrosanct and people will always value it in that way – that is a generational thing. I really think the Green Belt is something to be proud of as it is so well regarded, renowned and replicated across the world in different forms.

This shows the power of nostalgia, vision and imagination and discourse in planning although, clearly, there is the broader challenge of *learning* from the past whilst not allowing to prevent vision of an innovative future (Connell, 2009).

Discussion and Conclusion: Towards a More Multi-Scalar Spatial Approach towards Power in Planning

The chapter has weaved together the various frameworks by which power and politics in planning has been conceptualised and now returns to the crucial, overarching question of power in planning in relation to the Green Belt.

Which Groups Have the Most Power in Planning?

Whilst campaigners and some homeowners *attempt* to exercise power via Lukes' dimensions through planning's political nature, arguably a direct correlation cannot be drawn between these *attempts* and the *effective* wielding of power, with these agonist *attempts* regularly being circumscribed, modulated and 'shut down' by national Government/planning policy and the Planning Inspectorate (Parker *et al*, 2015, p. 519). Indeed, the power of campaigners is often fiercely contested by housebuilders and planners, sometimes through overt exercises of power, such as the appeals system, and other times more subtly by trying to sway public opinion on the grounds of sustainability or solving the housing crisis (Bradley, 2023; Lukes, 2004; Rydin, 1985).

More theoretically, whilst the Green Belt is *a* cause of the housing crisis, reality is more nuanced than it being, along neo-Marxian lines (Foglesong, 1986; Lake, 1993), the main weapon in home-owning capitalist armoury to defend high house prices. Arguably, even if the policy was abolished, there would still be opposition to development, high land/property prices and persisting, wider significant societal and generational inequalities regarding housing (which the Green Belt may *exacerbate* but is not the *primary cause* of) (Christophers, 2018, 2019; Ryan-Collins, 2018, 2021). Indeed, whilst planning is the *arena* in which competing groups, especially homeowners and housebuilders, *attempt* to exercise power and where conflicts are resolved (Short *et al*, 1987), it cannot be said, as some authors have implied (Airey and Doughty, 2020, p. 33), that the system generally or Green Belt in particular is 'rigged' or *disproportionately biased* in favour of one, 'rent-seeking' group, i.e. homeowners, as power is *fiercely contested* (see Figure 8.1). In this vein, the system as a whole probably does not need *radical* reform as was proposed by the White Paper neither does the Green Belt need abolition (Inch *et al*, 2020; MHCLG, 2020A).

How Is Power Exercised?

Lukes' (2004) *three dimensions* and the 'Politics' and 'politics' framework (Sims and Bossetti, 2016, p. 37), has illuminated overt attempts by campaigns to exercise

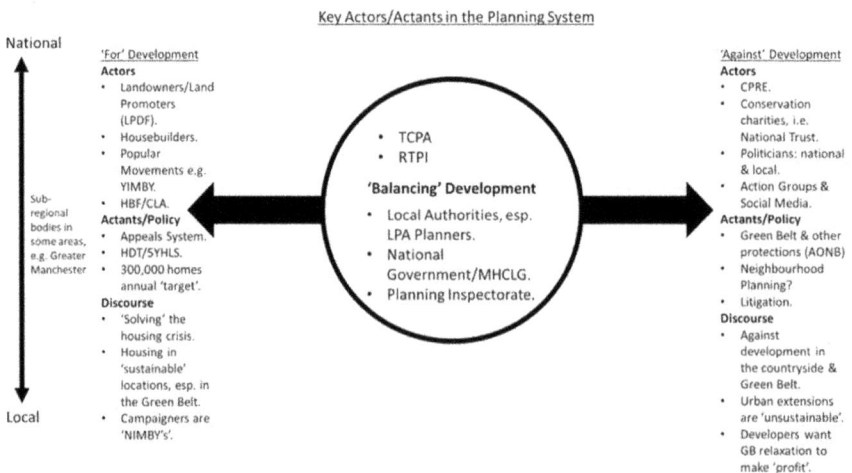

Figure 8.1 Conceptual diagram of key actors in the planning system (Author's Own).

power, such as a protest march around a proposed development site boundary in the Green Belt, alongside more subtle, discoursal attempts, like the 'politics of affect' associated with CPRE's campaigns about the countryside being 'concreted over' (Harrison and Clifford, 2016, p. 585).

The difficulties of empirically establishing the *effective exercise of power* by campaigners in this book probably reflect the central difficulty of defining what power 'is' and how effectively its exercise can be measured, which is particularly prescient given planning's conflicted and contested nature (Dowding, 2006; Flyvbjerg, 1998; Robinson, 2006). Operationally, a key limitation to Lukes' framework is the importance of space in the exercise of power and politics both in the decision-making process and power of discourse (Allmendinger and Haughton, 2012; Massey, 1994). Indeed, superficially similar places or LPAs can take different approaches towards Green Belt management and release, such as South Gloucestershire and North Somerset or Tandridge and Sevenoaks, due to *place specific factors and institutional culture* (Amati and Yokohari, 2006). This does not mean that Lukes' *three dimensions* cannot be used as a *basis* for theorising power in planning but suggests that multi-scalar factors are crucial with place-based empirical studies being vital for critically analysing how and why different actors and actants attempt to exercise power (Sturzaker, 2010).

On balance, the politicised nature of planning is perhaps inevitable and necessary to ensure that it has public transparency, democratic accountability and legitimacy (Goode, 2022A; Parker and Street, 2015; Raynsford Review, 2018B). However, societal needs from planning systems or them more effectively serving the 'public interest', especially through addressing climate change and delivering more affordable housing, has to be balanced against this need for democratic legitimacy (Lennon, 2020, p. 1; Tait, 2016, p. 335; Slade *et al*, 2019, p. 11). Arguably, the too-politicised governance structure associated with the localism agenda in England needs overhauling (hence the 'recommendations' Chapter (9) following this 'problematising' one).

Alongside the calculative turn in planning with the reflexive professionalisation of campaigning (Bradley, 2018, p. 24, 2021, p. 8), vision, nostalgia and imagination in planning continue to be vital for campaigners and planners. Clearly, planning involves evidence *and* vision but the system could begin to move away from some of its technical, numbers-driven nature to better harness popular vision and more productively involve the public in planning.

Broader Theoretical Reflections

This chapter has underlined the importance of knowledge production and categorisation in planning systems in which 'evidence' is crucial to decision-making. Whilst somewhat outdated and having significant limitations, including in defining the more philosophical issue of what power 'is' (which is beyond the scope of this chapter) (Dowding, 2006; Robinson, 2006), the chapter has evaluated the utility of Lukes' *three dimensions*, especially in terms of uncovering the modalities of power in planning systems including in the more subtle but prescient realm of discourse and construction of metanarratives. However, it has underscored the importance of the spatial and multi-scalar dimension to the exercise of power, which is crucially missing from Lukes' work. This reflects Allen's (2003, p. 102, 231) argument as to power being 'inherently spatial' and, drawing upon the *three dimensions of power*, the chapter has examined the modalities by which campaigners aim to establish multi-scalar relationships 'which establish their presence [and] imbue such institutional spaces with power' and how planning systems categorise their knowledge and modulates their influence at both local and national scales. Both Lukes (2004, p. 150) argument about assessments of power being always 'partial and limited' and Law's (2004, p. 795) point that 'surprisingly little can be said about [power's] operation in general' are accepted in that the theoretical conceptualisations in this chapter are based on empirical data focused on the English Green Belt and that more empirical studies are needed, particularly from the Global South (e.g. Roy, 2011).

The chapter has demonstrated both the need for more nuance in the literature regarding greater sensitivity to the 'gap' between the *attempted* and *effective* exercise of power by campaigners and the importance of the spatial and scalar dimension. This underscores Law's (2004, p. 795) argument that more case studies and 'empirical study is required' of power, especially in an international context. Indeed, beyond Rydin's (2020) study of the National Infrastructure Consent Process in relation to offshore wind and Bradley's (2018) work on neighbourhood planning, there is limited literature examining the specific modalities in planning, especially community opposition to new housebuilding.

In relation to scale, there has been wide-ranging debate in the literature regarding the definition, conceptual value and alternatives to hierarchal geographical 'scale' as a 'bedrock domain to ontology' (Marston *et al*, 2005, p. 416). This in turn relates to the geographical dimension of the exercise of power with power having been argued to be 'always already spatial' (Cidell, 2005, p. 231; Marston *et al*, 2005, p. 416). Some scholars, such as Brenner (2005, p. 9 cited in Marston *et al*, 2005, p. 416), continue to argue for the centrality and conceptual value of hierarchal scale whilst others, such as Marston *et al* (2005, p. 427), label it a 'scaffold imaginary' and argue for a 'flat' or 'multi-site' approach. Scholars such as Allen and Cochrane (2011) argue that conceptualisation needs to move beyond competing 'hierarchical' or 'flat' approaches to focus on the fluidity, topology and reach of power as there can be a 'multiplicity of scales…present at the same time' (see also Cidell (2005, p. 231)). Nevertheless, these debates demonstrate the importance of space in the exercise of power with Allen (2003, p. 89) drawing attention to both the 'hidden geographies' and 'quiet reach' of power. Empirical study is therefore needed to uncover these topological 'registers of power' (Allen, 2003, p. 154).

In the planning field, scale has been debated by scholars such as Allmendinger and Haughton (2012, p. 89) who argue that, using a post-political approach, the abolition of regional planning in England, for example, has been an attempt by Government to dissipate responsibility for challenging decisions through deliberately creating 'fuzzy' spaces of governance. However, other scholars such as Natarajan (2019) have highlighted the enduring importance of scale in planning with planning systems in many ways having hierarchal governance structures both in terms of policy and decision-making.

This chapter, through focusing on the Green Belt as a multi-scalar policy in England, therefore has examined the importance of geographical scale and hierarchy in the exercise of power as this has been largely neglected in Lukes' aspatial framework and is a significant gap in the literature (Allen, 2003). On the one hand, the chapter has found that power is both 'complex, multivalent, often conflicting and transversal' (Law, 2004, p. 795) and 'topological' (Allen, 2003, p. 102), especially in the spatial interconnectedness of both the decision-making process and construction of discourse and meta-narratives whereby campaigners attempt to extend their geographical 'reach' and influence to the national level in order to legitimise their claims to knowledge. Nonetheless, especially in an English context where regional planning has been abolished and the localism agenda is the dominant narrative, hierarchal, multi-scalar relationships remain prescient.

This is particularly the case in relation to the decision-making process, which is time-limited (Dobson and Parker, 2024), and where national Government extends significant reach and regulation through national policy and the appeals system etc. Indeed, in both a Foucauldian and agonist sense, there are arguably attempts by Government to silence dissenting voices and discipline dissident local authorities whilst, through a range of modalities, planners categorise and discredit campaigner 'knowledge' (Parker and Street, 2015, p. 794; Bradley, 2018). Integrating Lukes' *three dimensions* into a multi-scalar framework therefore enables conceptualisation of both the overt and subtle exercises of power at a range of spatial scales.

Of course, more research is needed into the characteristics of knowledge categorisation as an exercise of power in more zonal planning systems, where the scope for public involvement is probably less. Additionally, it is acknowledged that England has a distinctive context of intense politics regarding Green Belt and housebuilding both national and locally with this chapter charting how debates in planning are closely and reciprocally related to the political sphere. However, the chapter, alongside Chapter 7, has highlighted the broader 'public deficit explanation' whereby planners construct a 'conceptual binary' of campaigners as ignorant and not understanding planning systems (Welsh and Wynne, 2013, p. 552) as one reason for this process of categorisation by planners. Alongside this, the categorisation of knowledge is probably also related to the increasingly 'calculative' nature of many planning systems alongside the growing time pressure on decision-making (Dobson and Parker, 2024; McAllister, 2017).

In closing, the chapter has underscored the continued relevance of Lukes' *three dimensions framework* whilst highlighting the vital multi-scalar or spatial dimension as missing dimensions. Indeed, as recognising that power is 'inherently spatial' (Allen, 2003, p. 102), the chapter has conceptualised how hierarchal scalar power relations are still prescient in public policy with a range of actors operating at multiple scales – local, regional and national – trying to exercise and contest power in the production and categorisation of knowledge.

Notes

1 Local plans are often developed by planning officers in collaboration with councillors and, like planning applications, go out to public consultation (Brock, 2018; Goode, 2022A; Yarwood, 2002). At the LPA level, there is not a direct vote on planning policy as is often the case with zoning ordinances in the USA (Daniels, 2010). Communities can 'make' Neighbourhood Plans which can propose *minor* alterations to Green Belt boundaries – these go to a public Referendum (MHCLG, 2019, p. 40; Wargent and Parker, 2018, p. 390).

2 Interviewees identified North Somerset's rural character, its historical antagonism towards Bristol's growth and tradition of growth restraint as explaining its protective attitude, reflecting historical institutionalism and path dependency (Sorensen, 2015; Valler and Phelps, 2018). South Gloucestershire's less defensive attitude relates to it being a more mixed authority as rural *and* urban and how the Green Belt boundaries were initially drawn (see Chapter 9), despite also being historically dominated by the Conservatives.

3 I.e. councillors saying that they do not approve of decisions but have no alternative. In relation to Solihull's Local Plan (2020, p. 1), Cllr. Andy Mackiewicz, Cabinet Member for Planning, claimed that '15,000 homes may seem a lot, but we have legal duty…to meet Solihull's housing needs. This number is determined by national methodology. We do understand the concerns of residents but will keep Solihull as a great place to live'.

4 Cllr. Sue Cooper, Leader of South Oxfordshire Council, complained that the Housing Secretary's pressure to adopt the Plan or expect the County Council or Ministry of Housing to take control, was: '*An unacceptable intervention into local democracy…the Secretary of State removed the democratic right of South Oxfordshire's councillors and residents…*[we] *have now been unfairly silenced*' (Henley Standard, 2020, p. 1). The letter by Cllr. Peter Fleming, Leader of Sevenoaks Council, to the Planning Inspectorate, who 'failed' its Plan due to issues with DtC and housing numbers, was equally strong describing: '*The double down and attempt to bully us into withdrawing our plan. Well, we won't gamble on the future of our District or its environment and we certainly won't be bullied into withdrawing our plan*' (Flemming, 2019, p. 1). The Council's High Court challenge against the Inspectorate failed (Kahn, 2020).

5 The accuracy of assessing housing 'need' and household projections is also beyond this book's scope but the aim here is portray the views and strategies of campaigners on the subject.

6 The GL Hearn Greater Birmingham HMA Strategic Growth Study which was commissioned by the local authorities of the West Midlands conurbation.

7 In an unpublished article for CPRE Warwickshire (2019), he wrote that the 'housing numbers game' had taken an 'ugly turn'. The Government is *'moving the goalposts… any pretence that housing need will be objectively assessed using the latest available information has gone out the window…*(it) *is a political target pure and simple'*. (Quoted with permission from a copy of the article given to the author).

8 This Chapter does not evaluate the contentious and complex issue of how far these were campaigners 'real' concerns but focuses on their *techniques*.

9 How housing targets are calculated/allocated is beyond the book's scope but, although challenging, a robust, transparent and trustworthy method is clearly needed (Donnelly, 2019).

9 The geography and governance of the Green Belt

Introduction: Structures and Processes

This final empirical chapter draws together the preceding ones into recommendations and bridges the findings and conclusion. Many of the issues associated with the Green Belt in the data are governance-related, so this chapter examines recommendations by focusing on the policy's geography *and* governance at the strategic, regional/sub-regional scale. This is critically important because the Green Belt is an inherently *regional* growth management policy and strategic governance was a key area of consensus between campaigners and planners.

Drawing upon the regional spatial imaginary and historical institutionalist literature, this chapter conceptualises the Green Belt as an enduring, regionalising concept in the spatial vision of planners and professional campaigners which often contests and conflicts with its conceptualisation as a highly political, territorial and 'rationalistic' governance issue intensified during the era of localism (Mace, 2018, p. 4). It then evaluates the political 'feasibility' (Breheny, 1997, p. 210) of alternative governance arrangements for its strategic management through developing a broader, transferrable analytical framework of strategic planning which is cognisant of the current political context.

The chapter highlights the importance of strategic *structures* for effective decision-making throughout whilst acknowledging Harrison *et al's* (2021) argument that formal, statutory 'regional' planning covering a fixed geographical area, like the Regional Spatial Strategies, is unlikely to be revived due to a vastly changed political context and difficulties with predicting the future. It therefore focuses on *strategic* rather than 'regional' planning in governance terms whilst contending that the *concept* of the 'region' is still important in planning. Indeed, there is widespread consensus among planners/planning stakeholders on the need for some *form* of strategic planning of the Green Belt and housing numbers. Given the limitations to both 'hard' and 'soft' spaces of strategic planning, there is arguably more of a role for 'intermediate', fluid spaces of strategic planning in the future learning from networks like the West Midlands Forum and Standing Conferences (Goode, 2022B). Whilst the dismantling of strategic planning has been deeper in England than in other countries where there is sometimes a measure of statutory federal protection for regional planning, such

DOI: 10.4324/9781032674315-9

as Germany, the chapter highlights that vital, wider historic lessons regarding the longevity and durability of strategic planning can be drawn. Indeed, given the widespread practice of planning strategically internationally in federal states such as Australia and Switzerland, there is reflection upon what can be learned from the international literature whilst contributing to wider planning theory and practice.

Conceptualising Governance and the Green Belt

Green Belts as a Regionalising Concept

The broader literature and Chapter 3 have drawn attention to how certain policies or actors/actants can become institutionally embedded and acquire institutional characteristics which then 'frame' future trajectories or path dependencies (Martin, 2010; Sorensen, 2015; Valler *et al*, 2023). Moreover, the broader importance of vision and imagination in regional/sub-regional constructs and plans has been underlined historically as associated with key figures in the development of planning training and practice, especially Patrick Geddes and Patrick Abercrombie (Harrison *et al*, 2021). Building on the empirical material, this chapter explores how strategic vision and imagination continue to be vitally important in 'framing' strategic planning futures, thereby building on Valler *et al*'s (2023) work on the regional spatial imaginary of the Oxford-Cambridge Arc and institutional planning cultures in Oxfordshire. Indeed, the Green Belt can be conceptualised as a regionalising concept in normative terms among planners and *professional* campaigners (often retired planners), with the concepts of the 'region' and 'strategic', firmly embedded in their strategic vision and spatial imagination, especially from training at planning school (Valler and Phelps, 2018).

As the pre-eminent *regional* growth management policy in England and an internationally renowned policy, the Green Belt is a prime example of strategic thinking associated with Abercrombie and other planning 'heroes', such as Howard and Unwin (see Chapter 3; Amati and Yokohari, 2007, p. 312). It can also be conceptualised as a regionalising concept in the *popular* mind with work of Bradley (2019A, p. 181) drawing attention to the broader, regionalising vision that campaigners often have of the policy as an entity or object which connects a region together, like the M25 or M60. The chapter argues therefore that planners *and* campaigners often begin with a similar conceptual starting point or spatial imaginary of the need to view the Green Belt, housing and planning generally from a strategic perspective. This forms the empirical cornerstone on which recommendations for strategic governance are built.

Nonetheless, in (attempts) to resolve the central planning or governance 'dilemma' of protecting the Green Belt *and* meeting housing need, there is arguably a juxtaposition between the Green Belt as a regionalising, coordinating concept in the popular mind, policy terms and by planners/campaigners, and the fierce territoriality often displayed in the county system of governance (Wannop and Cherry, 1994, p. 52; Colomb and Tomaney, 2016; Riddell, 2020).

The Politics of Planning: The Green Belt as a Territorial Issue

However, the policy is often politically and popularly conceptualised as a more political, rationalistic issue with competing theorisations of it as a regionalising *and* territorial concept converging and competing in its governance[1]. For example, the rural shires, especially in the West Midlands such as 'fruitful' Worcestershire and 'leafy' Warwickshire, often laying claim and ownership to 'their' Green Belt to contain the 'industrial' conurbation (Hall *et al*, 1973B; Harrison and Clifford, 2016)[2]. Many of these governance issues are associated with the structure of local Government, especially administrative boundaries, which has resulted in often large, mainly Labour-controlled cities, like Birmingham, with limited room to 'grow' within their own boundaries surrounded by (Conservative or Liberal Democrat) rural 'shires' covered by the Green Belt (Hall *et al*, 1973A, pp. 617–622; Wannop and Cherry, 1994, p. 167). The institutions of the countryside, counties and Green Belt clearly intersect, but the regular construction of the Green Belt as a territorial issue has made and make building an enduring system of strategic Green Belt management in England a perpetual challenge.

However, as Chapter 3 outlined, Inch and Shepherd (2019) have explored the central importance of 'conjunctures' in planning history, and arguably, 2010 can be seen as a key conjunction with RSSs being abolished, the deepening housing crisis and diminishing supply of suitable brownfield land for housing resulting in the Green Belt becoming increasingly political and intensifying conflicts over the policy's governance (Inch and Shepherd, 2019). Although governance challenges and continual tension between cities and the counties have existed since the Green Belt was introduced, such as the Wythall Inquiry (1958), previously there were strategic *mechanisms* in place trying to resolve the planning 'dilemma', such as new towns (Lord and Tewdwr-Jones, 2018). Nonetheless, the dilemma has been particularly intensified since 2010 in areas *without* statutory strategic planning, especially the West Midlands.

Current Issues in the Governance and Geography of Green Belt: The Views of Planners and Campaigners

The Green Belt as a whole or strategic *entity* alongside historically being a regional growth management *policy* was explored by the planner interviewed from the RTPI:

> For most of the Green Belt's existence, there were satisfactorily strategic mechanisms in place so **it's only since 2010 there has been no strategic planning mechanism**... So it's the London Green Belt, not the Brentwood Green Belt... **If there is a need to reconsider Green Belt boundaries, it must happen at the level of the entire Green Belt, whether it is London or the West Midlands.**

Moreover, the lack of a strategic approach caused professional campaigners widespread concern regarding the Green Belt's spatial integrity because they viewed it as being released incrementally in a piecemeal way and 'nibbled away'. This was a particular concern in the West Midlands Green Belt where, without strategic planning, land from the Green Belt has been released for housing with campaigners giving examples of the Birmingham Development Policy, Warwick Local Plan (1470 hectares (of land)), Coventry Local Plan (1550 hectares accommodating 7000 homes) and the Draft Solihull Local Plan (Wilding, 2018)[3]. For example, Planner (1), CPRE West Midlands, argued:

> The original policy has become **seriously undermined**... really government policy now is driving a **coach and horses through the definition of what are exceptional circumstances**... It is not supposed to be ossified and last forever, which is not to say that you should go to the other extreme and feel you can change it at the drop of a hat, anytime you choose. There is a middle way which says, 'Let's have a strong Green Belt policy but let's be prepared to review the extent of the Green Belt every, I don't like to put a figure on it, **but say every thirty years'**. But we are so far away from that now.

Concerns were raised about the Green Belt's temporal permanence as this is nominally enshrined in national planning policy which underlines the need to '*keep land permanently open*' (DLUHC, 2023, p. 42). Nonetheless, the requirement to review local plans every 5 years fuelled campaigners' concerns about releasing land from the Green Belt for housing becoming the 'norm' and driven by short-termism in policy, especially in meeting housing numbers. Again, there was a chronic lack of trust with campaigners being anxious about Green Belt reviews being conducted largely by private consultants, which they viewed as 'biased' and 'incorrect' through assessing individual land parcels against the five purposes, such as the study of Hearn (2018), rather than viewing the Green Belt as a whole geographical 'entity'. Planner (2), CPRE West Midlands, argued:

> There should be a whole strategic review of it [WMGB] and not by these consultants employed by individual councils to review it and rank bits of parcels of land. They never consult the public and just bung it on a website!... it needs to be a really major public participation exercise and consultants wouldn't like that, because they are getting away with it.

Most private sector planners were also concerned and frustrated with the lack of strategic vision or management of the Green Belt. Their concerns centred on the failure of the Duty to Cooperate (DtC) to produce successful joint working between Local Planning Authorities. They also (largely) viewed locally led planning as fuelling parochialism among campaigners and reactive or defensive behaviour among LPAs as 'attempting' to lower housing numbers to minimise Green Belt release rather than proactively and positively planning for housing 'need'. A planner from housebuilder (1) argued:

> Bigger than local planning in the West Midlands is atrocious... there is a shortage in Birmingham of 38,000 homes and I see absolutely no evidence whatsoever of them (Councils) resolving the problem... it is very political and, when you had the regional agencies (RSSs), **the councils would say 'Well, you know, the figure has been imposed on us'.** That is why they were dissolved... But having now got a bottom-up system, we find that it is very politically uncomfortable for the local councillors... you are now in a situation of trying to deal with issues that are bigger than local but there is not necessarily any organisation to do it... it might be just one local authority looking at the Green Belt that is in their part/administrative area and, again, the whole thing is very political. So, there should be an overview of Green Belt and that might need to be done on a national or regional scale.

This quote is interesting because, although it shows a different perspective to campaigners regarding the necessity of Green Belt release, there is a united overall perspective of the Green Belt as a regionalising *concept*. Likewise, the Home Builders Federation Planner argued:

> In an old, old Plan [1964 South East Study]... They split the hinterland of London into areas and had various growth areas. The most famous one was Area 6 [Reading-Bracknell-Blackwater Valley-Basingstoke]... massive growth areas... because we knew that London was going to continue to grow economically and therefore we wanted to provide houses for people who had to move out of London because we were constrained [with the Green Belt]. There is [now] no regional planning... a strategic, (blanket) policy, like Green Belt, should surely be sorted out at a strategic level and **yet we don't have a strategic level of planning with which to do it**. So, it is left to people, like Birmingham City, to talk to Solihull, Bromsgrove and all the other surrounding authorities saying - 'Would you like to take some of our housing needs?' rather than saying - 'We have got a 60,000 figure to distribute, you are going to take 10,000 in a coherent sub-national plan'.

Popular, normative support for the policy as a principle seemed inextricably and reflexively linked, regionally and nationally, with rationalistic, political support for it as the HBF Planner also highlighted. The lack of strategic planning further politicising decision-making, especially contradictions in national policy and political interventions, like the Birmingham Development Plan, was highlighted, whilst Private Sector Planning Director (South East) (2) argued:

> You have a political choice, and which is more important - **protecting the Green Belt or putting development in the most sustainable place?** Strategic planning is something that the system needs... because, inevitably when

you are making big controversial decisions, as members going back to the politics in that district, you are all going to be affected to a greater or lesser extent by that decision and going to be lobbied by people who live in the district. It is very difficult to make difficult decisions. **Planning is as much about politics as it is about anything else.**

Interestingly, a national Conservative politician interviewed, who was sceptical of strategic planning, still recognised the challenges regarding the Green Belt's governance but justified the locally led approach:

It was never expected to be an easy ride [DtC] and sometimes it is difficult but I think it is the right approach because the alternative is that you automatically revert to a kind of national or a very high level of regional planning by bodies that don't have the local knowledge that LPAs have. So, I think it is the right approach and, of course, **the Green Belt is always one of the issues of greatest contention within that.**

However, planners and professional campaigners mostly perceive the Green Belt as a regionalising, coordinating concept (Bradley, 2019A), forming the empirical foundation for conceptualising feasible geographical alternatives.

Towards a Strategic Green Belt Approach and an Analytical Framework of Strategic Planning

Professional campaigners acknowledged that *if* the policy is to be reviewed, they would prefer it to be as part of a longer-term, strategic review[4]. In the West Midlands, CPRE and the Futures Network are taking a proactive role in advocating for strategic planning with the campaigner Jean Walters, MRTPI, setting out criteria for a potential Green Belt review:

- *It must be steered by a body without a vested interest in development.*
- *It must include representation by a wide range of interests – conservation and environmental bodies as well as developers.*
- *It must examine the strategic purpose of the Green Belt in question besides its geographical extent, for the two things go together.*
- *It must examine possible deletions from the Green Belt and possible additions to it on an equal footing. There must be equal potential for either change.*
- *A comprehensive review should not take place within 15-20 years of the previous review.*

Letter to Andy Street, Mayor of the West Midlands
Combined Authority (2018) (quoted with permission)

Although there was no universal agreement on the specifics of Walters' criteria among planners and campaigners, the four broad principles that she identified for Green Belt Review – temporal and spatial considerations, governance and

Table 9.1 Key criteria for strategic planning

Geographical scale	Decisions on strategic matters, such as Green Belts and locations of strategic housing allocations, need to be made at the 'larger-than-local' scale to enable strategic consideration of the *whole* Green Belt and allow *all* growth options, like additions/deletions and new towns, to be explored.
Timeframe	Long(er)-term plans are vital to add certainty to planning, give more predictability to developers, add more certainty to campaigners and reduce land speculation.
Governance	Ensure structures and decisions are beyond the *immediate* political cycle locally and nationally, so they are not dismantled after not being allowed to 'bed-in', like RSSs, and to reduce the amount of decisions made for short-term political reasons giving more certainty to the public/developers.
Legitimacy	The body needs to be seen as transparent and 'neutral' alongside having some democratic legitimacy to gain the trust and confidence of campaigners and the public.

legitimacy – strongly resonate with and are grounded in the quotes above so form the basis of an analytical framework to evaluate the feasibility of strategic planning (Table 9.1).

'Feasibility' is defined here as an institution's durability, with the RSSs being only short-lived, so this framework stresses the importance of governance and legitimacy, especially given planning's political nature (Breheny, 1997; Mace, 2018). The *process* of planning emerged in the data as vital alongside actual planning *outcomes* with public confidence in the system being key to its effective functioning (Parker *et al*, 2020). Moreover, most retired West Midlands planners highlighted the importance of professional *forums/networks* in which controversial issues, especially housing and Green Belt issues, can be discussed and seen *by the public* to be discussed (like Planning Committees or Regional Assemblies) (Goode, 2022B). Temporal and spatial considerations are very important and intertwined with Green Belt reviews having to cover a defined spatial area *and* particular timeframe, hence the framework's recommendations regarding time and space. Certainty is also vital for housebuilders, as the planner from housebuilder (2) argued:

So, it is basically lift it [the Green Belt] and review what you want for a longer period. Fifteen or five years as now in local planning is not enough, not when you are imposing something of this magnitude. Do that - be really ambitious but realistic.

The framework is therefore empirically based on a *regional* case study, but these broader principles have wider relevance and importance for strategic planning internationally, with some participants calling for better dissemination of 'best practice' regarding joint local plans and Green Belt reviews (Goode, 2022B).

Applying the Analytical Framework: Considering Alternative Governance Arrangements

Five main governance possibilities emerged from the data are as follows:

1 Grant planning powers to all combined authorities/strategic planning bodies to conduct Green Belt reviews, such as in the Greater Manchester Spatial Framework and the London Plan.
2 Reassemble voluntary, professional networks, such as the West Midlands Forum.
3 Create a new body or 'commission' for each Green Belt, such as national parks, to conduct a long(er)-term Green Belt review on behalf of LPAs. It would be composed of local or regional politicians alongside experts from developer, planner and campaigner backgrounds. It could be part of a broader, strategic plan. The commission would allocate areas of restraint/ development and review Green Belt boundaries so would have a clear remit of planning the Green Belt's overall land use for the long(er) term for each Green Belt. Strategic planning would focus on general strategic matters related to planning, such as transport and education, and cover the *whole* country.
4 Create 'intermediate' spaces of governance both in terms of process, like Green Growth Boards, and developing a flexible spatial blueprint such as integrated strategic frameworks.
5 Return to County Structure Plans, such as the recently aborted Oxfordshire Plan 2050, and is proposed in the 2024 Devolution White Paper (MHCLG, 2024).

Reviving RSSs was not included due to their abolition, electoral toxicity and probably not being feasible in governance or legitimacy terms (Allmendinger and Haughton, 2012). Each of these governance possibilities is now evaluated in turn based upon the data and analytical framework.

Combined Authorities

These are existing bodies and, being headed by a democratically elected mayor, give them democratic legitimacy. Crucially, they have political support with the Conservative local politician, Sutton Coldfield, arguing:

> The Mayor… is the guardian now of the economic interests of the West Midlands… he should seek to nail down where the housing numbers should be. It should be managed at the regional and sub-regional level and is probably best to be explored by the Mayor with the planning powers. **I want to give him the strategy and the planning authority can have tactics**… we need to build a lot more homes, but the critical thing is that, they must be in the right place. So, a Mayor is in a better place often to adjudicate where those places should be than the local authority.

However, there are still serious governance challenges because, where planning powers have been granted to combined authorities and strategic planning bodies, like Greater Manchester and the West of England, Green Belt release has still been very controversial and politicised, as subject to the electoral cycle, thereby undermining the longevity of Green Belt reviews (Bradley, 2019B; Branson, 2019). Moreover, for some combined authorities, especially the West Midlands, getting planning powers is too politically sensitive as these bodies rely upon voluntary co-operation (Young, 2020). Most of these bodies do not encompass the *whole* Green Belt so would have limited strategic oversight.

Return to County Structure Plans

Planner (2) from CPRE West Midlands advocated returning to (statutory) structure plans arguing that they allowed a more strategic perspective and gave the shires sufficient 'clout' to counterbalance, speak with one voice and 'stand up' to the cities. It is the preferred option of the former Conservative MP, Richard Bacon (Goode, 2022B), and was advocated by a County Councils Network (2020) report. However, due to austerity and difficulties in LPA resourcing, there is already an increase in 'joint' local plans within and across counties although many of these have run into difficulties such as the Greater Exeter Strategic Plan and Oxfordshire Plan 2050. Additionally, there is currently some restructuring in local Government towards unitary authorities, some at a county level[5] (Pike *et al*, 2018; Goode, 2022B; Riddell, 2020).

In governance terms, counties have democratic legitimacy, command popular support and structure planning works well where the county covers the county town or largest settlement and the surrounding hinterland, such as Oxfordshire and Herefordshire (Goode, 2022B). Nonetheless, county boundaries are not contiguous with England's current social and economic geography of city regions (Breheny, 1997). Furthermore, resurrecting them would potentially revive the perpetual tension and tussle between conurbations and counties over the Green Belt and be deeply enmeshed in the electoral system, potentially jeopardising the longevity of plans. Consequently, institutional memory, culture and path dependency can 'frame' the dominant future vision of place in counties, but this vision may not necessarily be the 'right' one in sustainability terms (Dühr, 2018; Valler and Phelps, 2018, p. 699)[6].

Reassemble the Voluntary, Professional Networks

Some retired planners advocated reviving strategic planning by voluntary networks and their professional expertise, such as the regional planning officers group:

> Local authorities don't come together anymore... [like] the West Midlands Forum and this is before Regional Assemblies. The Forum came out of quite an early form of regional planning in the late 60s, early 70s and it was a

non-statutory body, so it was a voluntary grouping, but they felt this was a way of mutually discussing the region's planning issues.

(Retired Structure Planner (West Midlands))

Reassembling voluntary networks would be more straightforward to achieve as being voluntary, not statutory, some interviewees argued that more can be achieved when relying on mutual goodwill and trust. However, this raises the question as to *why* Governments have not reassembled them. Additionally, there are issues with democratic legitimacy when experts are perceived to be making decisions outside of the public's view and 'behind closed doors' (Allmendinger and Haughton, 2012). Furthermore, as with the DtC, some governance issues, like the Green Belt and housing numbers, are so controversial that they require statutory steer and underpinning, as a private sector young planner (national) argued:

It's for national Government to make a change for the better... in terms of coming up with a structure, whatever that might be, for regionally working out what to do in the South East... it's not going to be local authorities or even Mayors - it needs to be nationally.

Creation of a Green Belt Commission

The proposal of 'the Commission' was developed as a conceptual ideal by the author based on the analytical framework and interview findings. As with national parks (Maidment, 2016), quasi-political and expert-led governance models can be effective and long-lasting and generally command the confidence of campaigners, developers and the public. The Commission would give the unique governance possibility of covering the *whole* Green Belt *and* taking strategic decisions for the long term, perhaps 25–30 years, on which land should be protected and developed as part of a strategic sub-regional/regional plan, such as the 1964 South East Plan. Indeed, the idea of a Single Joint Expert Group to make decisions on the Green Belt has been advocated by Christopher Young (2019), a leading planning King's Counsel. However, Young envisaged that this would be led by the private sector, while this chapter argues for a more mixed or balanced body made up, such as national parks, of locally elected politicians alongside planning experts to help ensure democratic legitimacy (Table 9.2).

Private sector young planner (national) also underlined the importance of such a body covering the *whole* Green Belt:

There needs to be a body which reflects the area that needs to be reviewed... So, it is how you go about doing that when the Metropolitan Green Belt extends so far outside London. I think realistically (and also to be effective), **any strategic body needs to take that really broad strategic view of the whole of the area that it could encompass.**

Table 9.2 Towards a Green Belt Commission: Initial recommendations

Geographical scale	A commission would be in each major Green Belt, sitting alongside a strategic plan, and would review the *whole* Green Belt, allocate areas of restraint/growth and draw boundaries. This would follow a national plan which has reviewed the Green Belt's purpose and *overall* spatial extent.
Time frame	Ideally 25–30 years, probably 15–20 years, to give long(er)-term certainty around Green Belt boundaries to landowners, housebuilders and campaigners and build public confidence.
Governance	Formed of experts and locally elected politicians making decisions, like national parks. A clear remit of a long-term review rather than deciding individual planning applications.
Legitimacy	Being formed of local politicians alongside experts would potentially build public confidence surrounding its objectivity, compared to private planning consultants largely conducting Green Belt reviews as now. Like a Royal Commission, it would take representations and evidence from interested parties.

Of course, such a body could be problematic, especially regarding democratic legitimacy, as Public Sector Planner (3) West Midlands explained:

If you had a Green Belt Council, I suppose the obvious issue with that is where does it get steer from? It is probably introducing more uncertainty.

Herein lies another central governance 'dilemma', especially the tension between democratic legitimacy and the necessity of long-term strategic decision-making based on evidence (Wannop and Cherry, 1994, p. 52; Sturzaker, 2017; Raynsford Review, 2018B). A Commission which is less politicised as beyond the immediate electoral cycle but still includes politicians in decision-making is a balanced approach, especially as frequent changes in council control under the current system are what often create uncertainty and render decision-making difficult, such as South Oxfordshire (Goode, 2022A). Indeed, the inadequacy of localist, combined authorities and other approaches, such as structure planning, means that it is important to consider with more vision about whether strategic planning and the policy could be managed more successfully through 'the Commission' as the approach most effectively incorporating various aspects of the framework.

'Intermediate' Spaces of Governance

There are the political challenges of 'hard' spaces of strategic planning, such as the RSS, in terms of democratic legitimacy and the potential rigidity of a fixed spatial blueprint, such as Abercrombie's *Greater London Plan* or *Conurbation*, at a time of tremendous economic and social change (Harrison *et al*, 2021). However, the profound governance challenges since 2010, especially in the West Midlands, also demonstrate the complexity of, and limitations to, 'softer', voluntaristic or more 'fuzzy' forms of governance (Boddy and Hickman, 2020). Based upon the findings of this chapter about the limitations to both 'hard' and 'soft' spaces of strategic planning, in the short-medium term, there is arguably more of a role for 'intermediate', fluid spaces of strategic planning moving forward learning from networks such as the West Midlands Forum and Standing Conferences.

As was done before 2004, national Government could perform a vital function in setting the overall structure and process of mandating key actors to work together strategically through overarching regional planning guidance, but there could be less prescription as to *modes* of governance or the level of detail that strategic plans cover. This would be learned from the key reflection of RSSs being critiqued as too 'unwieldy' (Allmendinger and Haughton, 2011; Valler *et al*, 2023). This is the direction of travel of policy and practitioner debate in England with the RTPI arguing for strategic Green Growth Boards to bring together key actors, and the County Councils Network has advocated for integrated strategic frameworks (sub-regional, non-statutory strategic plans) and strategic planning advisory boards (to enable strategic discussions among the main sub-regional players). Moreover, the Government has set out its aspirations of strategic planning through Devolution White Paper (MHCLG, 2024, p. 14), which requires "all areas, with or without a Strategic Authority, ... to produce a Spatial Development Strategy".

Discussion and Conclusions

This chapter has addressed the fundamental issue in this book that planning is often conflictual, controversial and contentious as balancing different, often competing, interests (Rydin, 2011, 2013). It has focused on the central governance 'dilemma' of meeting housing needs *and* protecting the environment (Short *et al*, 1987, p. 40). As the strongest protection against development, this dilemma is often most challenging regarding the Green Belt with the chapter examining how the localism agenda has exacerbated the dilemma and exploring the governance *process* and *spatial scale* at which it should be resolved.

The chapter has addressed many of the issues raised in Chapter 3 about the importance of the Industrial Revolution, which has created an enduring dualism of rural and urban, especially in the West Midlands. Indeed, the Green Belt is popularly seen as personifying one of England's greatest institutions – the countryside (Mace, 2018). As Chapter 8 and this chapter have underlined, this has become inscribed in the country's political geography with the widespread territorial desire of often Conservative counties to preserve 'their' countryside and Green Belt from encroaching 'industrial', usually Labour, conurbations (Goode, 2022B).

This shows how similar institutions, including the Green Belt and counties, can overlap, intersect or militate against each other although, more broadly, these institutions are tied up in nationalism with the popular contrasting of the *English* system of counties with the 'other' *technocratic* European approach of 'regions' (Colomb and Tomaney, 2016). The power of normative, cultural and historical institutionalism can be seen in the durability of counties and the Green Belt notwithstanding the growing importance of the concept of city regions internationally, such as Grand

Paris, alongside the requirements of capital and neoliberal logic for strategic planning (Dembski *et al*, 2021).

Although the Green Belt has consistently been an inherently political issue and strategic planning has had a troubled history in England due to these deep historical factors, in a 'hyper' political era of the localism agenda, Brexit and Trump (Barry *et al*, 2018; Jessop, 2018), this chapter has argued that the Green Belt (and planning systems more broadly) is becoming increasingly politicised, notwithstanding scholars' arguments on post-politics (Allmendinger and Haughton, 2011, p. 314). Indeed, the strategic planning 'agenda' has arguably suffered between the 'rock' of increased Government centralisation of planning (see Chapter 8) and the 'hard place' of the localism agenda (Boddy and Hickman, 2018, p. 198; Lord and Tewdwr-Jones, 2018). Furthermore, as outlined in Chapters 3, 4 and 7, this political geography is interwoven with social geography in England of widespread opposition to housebuilding motivated by fear of change surrounding the popular planning principles of the countryside and Green Belt alongside place attachment.

However, whilst this historical and current context poses great challenges for planners and scholars in making the 'case' for strategic planning and the prospects of the federal Government may currently appear as a theoretical ideal in England (Harrison *et al*, 2021; Rae, 2016)[7], the demise of statutory regional planning has failed to 'solve' the pressing and deepening housing crisis, even if measured on the narrow metric of housing supply (as documented in Chapters 3 and 6) (Bradley, 2023; Inch and Shepherd, 2019; Lund, 2017). However, the electoral success of political parties depends upon successfully 'solving' this housing crisis to widen homeownership (Inch and Shepherd, 2019; Tait and Inch, 2016).

Moreover, the discontent with locally led planning among planners and campaigners, which this chapter has documented, has arguably found a broader popular political backlash, especially against the frequent Green Belt releases of the localism era, as seen in the 2019 local elections and 2021 Amersham and Chesham by-election. Wannop and Cherry (1994) argued that the consensus underpinning post-war strategic planning was that development locations need to be planned strategically, and crucially, because planning decisions are often difficult and controversial, local politicians needed to be able to deflect and redirect the 'blame' of strategic decisions to the regional or national level. Consequently, given this governmental logic and proposals by the Labour Government for a 'strategic approach' towards Green Belt and planning more broadly articulated in the Devolution White Paper (Martin, 2024; MHCLG, 2024), it now appears that there *is* widespread political acceptance that some *form* of strategic planning is needed, so it is vitally important for planners and scholars to critically consider alternative *forms* of strategic planning (as Wargent and Parker (2018) did for neighbourhood planning). This chapter has begun this process for the Green Belt but also for strategic planning more broadly by developing an analytical framework.

However, until the 2024 election, the Government appeared to pin its hopes on voluntary Devolution Deals, city regions alongside joint local plans although the centralised nature of these 'deals', the power imbalances and 'strings' that they entail, geographical unevenness in their take-up and neoliberal nature of their governmentality have all been critically outlined by academics (McGuinness and Mawson, 2017)[8]. The future of strategic planning was challenging as comprehensive local Government reorganisation towards county unitaries in the repeatedly delayed Devolution White Paper, which changed into the Levelling Up White Paper, did not go ahead although this is now proposed in (Labour's) Devolution White Paper which also restores the requirement for strategic planning (County Councils Network, 2020; Riddell, 2020; MHCLG, 2024). The move towards 'voluntary' housing targets and optional Green Belt reviews in 2023 did not provide a context very conducive towards supporting strategic planning although these changes have now been reversed by the Labour Government with mandatory housing targets restored (Young, 2023; Ministry of Housing, Communities and Local Government, 2024).

Nevertheless, whilst the Planning *White Paper* was shelved, the Green Belt was one of the few parts of the planning system that the neoliberal Conservative Party did not propose completely overhauling. Moreover, whilst this chapter has underlined the importance of strategic planning in *governance* terms, it has also argued that the 'region' and 'strategic' is an enduring and prescient *concept* for planning and geographical theory and practice. Indeed, the chapter has demonstrated how the region remains important even during the localism agenda and a hostile institutional context in England with the concept having particular poignancy in countries with federal systems of governance like Germany and the USA. However, the Green Belt is *the* pre-eminent example of the enduring importance of the region as a regional growth management policy, and it is still conceptualised as a regionalising, coordinating concept by planners *and* professional campaigners. Based on this theorisation and the governmental logic for strategic planning, the chapter evaluated alternative governance arrangements for strategically managing the Green Belt, based on the case study of the West Midlands.

However, a strategic Green Belt review would depend on being part of a broader strategic plan, so the chapter has also focused on the importance of strategic planning generally. This would enable strategic consideration of the various spatial visions and growth/restraint options available, such as new towns, *alongside* the Commission's review to try to more proactively and productively involve the public in planning (see Chapter 8). It would also sit within and follow a national plan reviewing the overall purpose and spatial extent of the Green Belt (see Chapter 6) and a national Green Belt conversation (Chapter 7). Nevertheless, in the short-medium term, there is arguably more of a role for 'intermediate', fluid spaces of strategic planning. Reviving 'intermediate' spaces of governance would rise to Harrison *et al*'s (2021) central challenge of how planning for 'regional futures' can be revived and sustained without the resurrection of formal, statutory 'regional planning'. Intermediate spaces of governance have broader potential given the issues with strategic planning in many nation, particularly the reluctance of multi-scalar actors to work together and the complicated constellation of actors involved (Dühr, 2018).

Finally, this book's key aim has been not just to be critical of the current system but also to suggest viable alternatives (Campbell *et al*, 2014; Fainstein, 2010). The empirical chapters have worked through the various aspects of Breheny's (1997, p. 210) 'feasibility test' with Chapter 6 dealing with the practical and economic feasibility of the extent to which the Green Belt needs reforming to solve the housing crisis, Chapter 7 exploring the social acceptability of Green Belt reform and Chapters 8 and 9 exploring and addressing the central premise of this book – the political nature of planning, especially the Green Belt. However, it is ultimately up to politicians to decide how they would like to manage the Green Belt although this chapter and book have sought to make a strong case for strategic planning.

Notes

1 'Rationalistic' is used here in a political sense with politicians often using the policy as a tool to gain and retain power (see Chapter 7; Mace, 2018, p. 4).
2 The Herbert Commission (1960, p. 186) referred to Surrey County Council extending the Metropolitan Green Belt to ensure that: *'If London's population overlaps the Green Belt, as it's clearly doing, the emigrants shall alight, say in Hampshire or Sussex, rather than in Surrey'*. Self (1962, p. xx), reflected after the Wythall Inquiry that: *'The Government warmly recommended the implementation of restrictive Green Belts but held aloof from the complex problems of urban dispersal which must be solved if Green Belts are workable'*. This is a particular issue in the West Midlands with the historic antipathy of the counties towards Birmingham due to its rapid growth and worries about the city's boundaries expanding with its territorial 'assertiveness' (surrounding settlements were 'annexed' in 1909/1911/1928/1932/1965/1974) (see Chapter 3; Cherry, 1996, p. 152).
3 A retired planner interviewed from Avon County Council highlighted 'pink areas' in the Avon Green Belt's (AGB) map from the 1975 Cribbs Causeway Public Enquiry. These were 'safeguarded for future use' and subsequently used. The author explored the *Diaries of Richard Crossman* (1975, p. 65), the housing minister, and the entry on 19 February 1965 read: 'I approved my Local Government Boundary Order for Bath and Bristol, the first really controversial decision I have taken'. Due to the minister's foresight in insisting on 'pink' areas, there has been less development pressure on the AGB than the WMGB with its tightly drawn boundaries. Private Sector Planning Director (South East) (2) argued that as most of the 'pink areas' have been developed, it is time for a review of the AGB.
4 At the London Plan Examination, which the author attended, Richard Knox-Johnston, Chairman of CPRE Kent, argued that if a review took place, it should be of the *whole* Metropolitan Green Belt.
5 For example, the Oxfordshire Plan 2050 covered the county (Private Sector Planning Director (South East) (2)).
6 For example, Oxfordshire's dispersed growth and county towns policy where jobs have continued to grow and centralise in Oxford, but there is insufficient or slow public transport, especially from Witney via the A40. This results in notorious traffic congestion (Dorling, 2019, communication between Dorling and the author).
7 The power of central Government and lack of federalism, like Germany, mean that regional layers of governance can be swiftly abolished, e.g. Greater London Council/ RSSs (Allmendinger and Haughton, 2012).
8 Devolution or city deals are voluntary arrangements agreed between councils and the Government to work together in exchange for funding and increased devolved functions, sometimes including strategic planning (Lowndes and Pratchett, 2012). They sometimes cover city regions, like Greater Manchester.

10 Overarching implications for planning theory and practice

Introduction

This conclusion chapter is the culmination of the empirical chapters on housing (Chapter 6) ((on) policy), community opposition (Chapter 7) (the public and protesters), politics (Chapter 8) (politicians) and governance (Chapter 9) (process) which, in turn, were based upon the chapters on researching the Green Belt (Chapter 2), conceptual framing and the history and form/function/evaluation of the Green Belt (Chapters 3–5).

Although each chapter drew out theoretical and policy implications, these are broadened here by synthesising the book's wider implications for planning and geographical theory, policy and the planning profession. The chapter argues for more integrated, positive strategic planning, as takes place in other countries such as the Netherlands, Sweden and Germany, and develops the study's overall, broader theoretical implications, especially the central challenge of reconciling community opposition with strategic planning, the past and present with the future and public opinion with the 'public interest', particularly the national (or regional) interest with local interest(s) (Tait, 2016, p. 335). The role of planning systems and planners balancing a wide range of interests and different, often competing, knowledge is stressed alongside making the case for the academy and practice to strengthen their mutual links (Allmendinger and Haughton, 2012, p. 90; McDowell, 2016, p. 2093; Raynsford Review, 2018B; Bicquelet-Lock, 2019). Areas for further research are highlighted before offering final reflections.

Drawing upon a mixed-methods approach, the book evaluated the extent of the contribution of the Green Belt to the housing crisis and assessed how far it needs to be reformed. It conceptualised the policy as a regionalising, coordinating concept alongside theorising more broadly about community opposition to housebuilding and politics and power in planning systems. In this way, the book focused on both the applied and theoretical aspects of the policy and wider planning systems and examined both the *process* and *outcomes* of planning which have traditionally been explored separately in research (Rydin, 1985).

DOI: 10.4324/9781032674315-10

Contributions to Theory and Understanding of the Changing Context for Planning Practice

The Contribution of the Theoretical Framework Developed in the Book

Regarding the central theoretical question about power in planning systems, the book has identified a prescient 'gap' between attempted and effective exercises of power with these exercises being contested at a range of spatial scales. Crucially, whilst acknowledging temporal fluidity in constellations of power, the book has highlighted the vital multi-scalar dimension as missing from Lukes' *three dimensions*. It has examined the modalities by which campaigners aim to establish multi-scalar relationships 'which establish their presence [and] imbue such institutional spaces with power' and how planning systems categorise their knowledge and modulate the influence of campaigners at both local and national scales (Allen, 2003, p. 231). There is, therefore, resonance in these findings with Allen's (2003, p. 102) work on the importance of space and spatiality to exercises of power with power being 'inherently spatial'. Nonetheless, whilst acknowledging the fluidity and topology of power (see also Allen, 2003; Allen and Cochrane, 2010), the book argues that hierarchal scalar power relations are still prescient in public policy.

It has underscored the enduring usefulness of Lukes' *three dimensions* in highlighting overt and subtle attempts at exercising power whilst acknowledging that the concept and effective exercise of power is difficult to define as often fluid and contested (see Chapters 5 and 8) (Dowding, 2006). Indeed, the literature on power, especially power in planning systems, is probably not sufficiently nuanced, and this raises deeper philosophical questions about what power 'is' and makes theorising power in planning systems more challenging (Matthews *et al*, 2015; Robinson, 2006). This demonstrates Cherry's (1982, pp. 116, 117) observation about power in Marxian theory – 'explanations of town planning are much more difficult... more penetrating analysis suggests that the reality is much more muddied'. Furthermore, Lukes' (2004, p. 150) argument about assessments of power being always 'partial and limited' and Law's (2004, p. 795) point that 'surprisingly little can be said about [power's] operation in general' are accepted as the theoretical conceptualisations in this book are based on empirical data focused on the English Green Belt, and more empirical studies are needed, particularly from the Global South (e.g. Roy, 2011).

Nevertheless, whilst the policy is distinctive, there are similarities to community opposition to housebuilding in the Groene Hart (Green Heart) in the Netherlands, Golden Horseshoe in Ontario and the Urban Growth Boundary in Portland (Oregon) (Altes, 2017; Zonneveld, 2007). Indeed, through focusing on the modalities of power through Lukes' framework, the book has demonstrated both the need for more nuances in the literature regarding greater sensitivity to the 'gap' between the attempted and effective exercise of power by campaigners and the importance of the spatial and scalar dimension. This underscores Law's (2004, p. 795) argument that more case studies and that 'empirical study is required' of power, especially in an international context.

Theorising the Changing Role of the Green Belt

Compared to most countries which do not have a rigid regional or national growth management policy, a central juxtaposition can be seen in the Green Belt's continuing presence, popularity and prescience as a planning policy notwithstanding the deepening deregulation of the planning system and broader neoliberalisation of the British economy (Mace, 2018; Prior and Raemaekers, 2007). The book has underlined how this can be explained both through the 'deep' historical factors of the Industrial Revolution and popular romanticisation of the English countryside *and* the adaptability and flexibility with which the Green Belt's purpose has evolved over time.

For example, the primary role popularly conceptualised for the policy in the inter-war period was for recreation changing to preventing urban sprawl and protecting agricultural land in the post-war era and then evolving to supporting the redevelopment of brownfield land for housing, urban regeneration and ensuring compact, 'sustainable' development to help address climate change (Amati and Taylor, 2010, p. 143). The primary role of the Green Belt is arguably experiencing another evolution currently with the importance of 'rewilding' and green infrastructure coming to the fore with the Environment Act, the post-Brexit subsidy regime, climate change and post-COVID recreational access to the countryside (Kirby and Scott, 2023; Kirby *et al*, 2023A, 2023B and 2024; HM Government, 2018; Goode, 2022A). This underlines the Green Belt's particular adaptability, durability and robustness as an institution, resonating with the popular desire to protect the English countryside and reflecting how its primary purpose evolves to reflect the key changing priorities of planning policy. However, whilst to some extent this reflects the distinctiveness of English popular culture, the Green Belt demonstrates both the robustness and adaptability of institutions in their historical evolution, thereby underlying the conceptual value of 'historical institutionalism' in theorising long-standing public policies and institutions, like the NHS (Valler and Phelps, 2018, p. 1 and Sorensen, 2015, p. 15).

Theorising the Changing Multi-level Politics of English Planning

In many ways, therefore, the politics of the Green Belt, especially its governance, reflects the broader shifts in the political landscape, whilst it retains the timeless overall political and popular appeal of preserving the countryside and preventing urban sprawl. Nevertheless, whilst the waxing and waning of regional planning can clearly be seen historically before 2010 and there were evident ideological attacks on planning, the period since 2010 can be seen as a key 'conjuncture' and distinctive planning 'moment' in politics with the dismantling of strategic planning, dominance of the localism agenda and increasing central Government intervention in the planning system (Inch and Shepherd, 2019, p. 2; Lord and Tewdwr-Jones, 2018). This has produced the central juxtaposition whereby the Green Belt remains a nominally strategic growth management policy yet lacks strategic governance arrangements to effectively manage it (Goode, 2022B). Indeed, the UK remains distinctive internationally in the current lack of strategic planning arrangements as it is not given federal protection like in other countries, such as Germany and the USA. Whilst there are signs of political backlash locally and nationally against the local release of land in the Green Belt for housing, which has characterised the era of localism (see Chapters 8 and 9), the broader political landscape arguably has been significantly changed since 2010 as remaining hostile to planning, especially regional planning, although this changed with the 2024 Election (MHCLG, 2024).

However, the practical operation of the Labour Government's approach, including the promised 'strategic approach' towards the Green Belt, remains to be seen (Young, 2021; Goode, 2022B; Martin, 2024, p. 1). This reflects broader critiques of regional planning and Harrison *et al*'s (2021, p. 1) provocation that 'regional planning is dead', so this book has explored the potentiality and feasibility of more 'intermediate' spaces of governance (see Chapter 9) (Valler and Phelps, 2018, p. 1).

Theorising the Importance of Space and Scale

The book has drawn extensively upon the planning and geographical literature both in initially theorising the Green Belt and subsequent data interpretation, especially on geographical and governance issues, showing the importance of these literatures interchanging and 'speaking' to each other (Allmendinger and Haughton, 2012, p. 90). In contrast to the aspatial nature of much of the economic literature and the frequently 'place neutral' and 'policy blind' nature of national policy (McGuinness *et al*, 2018, p. 330), the Green Belt *is* a prescient, inherently *geographical issue* with a spatial purpose, preventing urban sprawl, so geographical considerations *must* be a vitally important component in any reform (Goode, 2022B). Indeed, the policy demonstrates the importance of geographical variation in that, whilst it is a national policy, issues get *refracted* or intensified by local circumstances underlining the importance of place specificity in the exercise of power (Allen, 2003; Amati, 2007).

Whilst somewhat reflecting the specificities of the localised nature of the English planning system, this demonstrates the importance of spatial interconnectedness, especially the interconnectedness of the local and national. Indeed, national politics and policy shape local campaigns, but campaigners seek to 'scale up' and generalise their claims to knowledge in order to appear legitimate (Cahill, 2004; Roy, 2011). Indeed, whilst there has been debate about the relevance and validity of 'scale' as a concept in human geography (Marston *et al*, 2005), it arguably continues to be of prescience in the realm of public policy, especially planning systems, which are characterised by subsidiarity and hierarchal (spatial) policy scales in decision-making (Natarajan, 2019). For example, although largely managed by *local* authorities with the current governance vacuum in strategic planning, the Green Belt is a *national* planning policy with central Government scripting and setting the parameters within which the policy is managed. Likewise, as seen in the 2024 general election campaign, the Green Belt is a national as well as a local political issue (Martin, 2024). The housing crisis is also a multi-scalar phenomenon as shaped by both international flows of capital *and* locational characteristics, whilst housing policy is largely determined nationally in the UK (Bradley, 2023; McGuinness *et al*, 2018, p. 330).

Another key debate in geography and planning is the relationship between the global and local and, in particular, how generalisable findings and theory are from qualitative approaches, such as ethnography, feminist geography and participatory action research, when these approaches often stress the particularity of place (McDowell, 2016; Nagar *et al*, 2009; Peck, 2015). Some scholars, like Cahill (2004), have attempted to discuss global, megatrends at the local scale, such as gentrification with women in New York's neighbourhoods. In the field of planning scholarship, the debate around scale has focused on community involvement and neighbourhood planning (for example, Wargent and Parker (2018) and Natarajan (2019)). The Green Belt is *a* poignant multi-scalar issue and policy as a *national policy largely managed locally*. Consequently, multi-scalar planning policies are very fruitful, albeit under-researched, objects of study to theorise planning systems and wider society as this book has sought to do.

For example, researching the Green Belt and its relationship with the housing crisis in this book has opened up the possibility of exploring how it has shaped wider urban development patterns in England compared to other countries. Interrogating the multi-faceted, multi-scalar nature of the housing crisis and reflecting on international comparisons have informed recommendations on the Green Belt's reform, especially the importance of affordability (Chapter 6) (Christophers, 2018, 2019; Munro, 2018). Focusing on the motivations for communities opposing development opened a rich seam of national/local and principle/place attachment factors shaping and forming campaigners' views (Chapter 7). Examining the politics and governance of the Green Belt reveals the close nexus between national and local Government in England, which is symptomatic of its centralised state, and opens up the prescient questions about the constitution, federalism and local Government reform (Chapters 8 and 9) (Lord and Tewdwr-Jones, 2018). This shows the profound importance of space as a very significant shaper of societal relations, whilst these relations are inscribed into and can be interpreted from the contours of space (Castree, 2008; Massey, 1994). Geographical theory is therefore very helpful in conceptualising the Green Belt, whilst this book has found that a great deal can be 'learnt' in theorising from the policy regarding broad questions, such as power in planning and society.

The Green Belt and the Housing Crisis

The Housing Crisis as a Multi-faceted, Complex 'Wicked' Problem

In operationalising theory and answering the central research question about the housing crisis, which helps empirically elucidate power in planning systems, the study found the housing crisis to be a multi-faceted, complex conjuncture of local, national and international trends rather than *purely* about the Green Belt and planning regulation (Bradley, 2023; Hudson and Green, 2017; Pike *et al*, 2018; Wetzstein, 2017). In particular, the book underlined the importance of locational characteristics which have poignant implications for policy as many countries look to address their housing crises often through planning reform. This underlines the need for more interdisciplinarity in housing research, but planning theory is ideally placed to utilise these various theories being at the confluence of multiple disciplines (Ormerod and MacLeod, 2019).

The Relationship between the Green Belt and the Housebuilding Industry

Most planners, including in the private sector, were supportive of the *concept* of the Green Belt arguing that the policy provides certainty and stability for the development industry by preventing the market from being 'flooded' with new housing, thereby 'crashing' house prices as to some extent happened in Ireland and Spain before the Financial Crisis (Kilroy, 2017). This suggests that housebuilders, planners and the development sector largely take a more nuanced view towards planning regulation than the largely critical position that the literature presumes (Raco *et al*, 2019). Indeed, this book found that regulation is often important for underpinning stability and certainty in a (development) sector which is characterised by risk and volatility (Raco *et al*, 2018, 2019). Moreover, the Green Belt to some extent perpetuates the oligopolistic power of the volume of housebuilders in the land market, and although this is a particular feature in England, housebuilders have significant influence in many planning systems internationally (O'Callaghan and McGuirk, 2021). This also relates to the broader international literature on the impact of planning regulation on the size of housebuilders and planning consultancies although more research is needed in this area, especially in relation to opaque land markets (Ball *et al*, 2014).

The Relationship between Green Belt Reform and the Housing Crisis

Nonetheless, the Green Belt still contributes towards the housing crisis, and some reform is needed, especially as a significant change in the broader economic structure of homeownership and volume housebuilders constructing most new housing is unlikely anytime soon. The Green Belt often *exacerbates the housing crisis in particular locations*, especially in conurbations, like Birmingham, with limited room to accommodate housing 'needs' within their own boundaries on brownfield land, particularly 'family' homes with gardens. Notwithstanding initial author expectations of this being largely a Metropolitan Green Belt problem, it kept reoccurring as an issue around the country, both in fast-growing cities, such as Bristol and Oxford, *and* in post-industrial cities, especially the West Midlands and Greater Manchester, thereby underlining the importance of geographical, case study–based research of the policy. The book has therefore proposed a national review of the Green Belt's overall purpose and spatial extent (Chapter 6).

Nonetheless, abolishing the policy *itself* is unlikely to solve the housing crisis, thereby further weakening the central research proposition that it *disproportionately* benefits, and is *primarily* supported by homeowners as increasing house prices (Chapters 2, 4 and 6). This highlights the utility of using a social justice framework of analysing who 'gains' and 'loses' from the Green Belt as a way to evaluate public policy (Kiernan, 1983, p. 83; Flyvbjerg, 2004, p. 292). Indeed, the book found that its social impacts are more nuanced than Hall *et al*'s arguments (1973B, p. 433) in *Containment* and that it is *not* the *direct* result of effective 'rent-seeking' by homeowners as there are important safeguards in planning systems that check the exercise of power (see Chapter 8; Ball *et al*, 2014, p. 3010). More fundamentally, this demonstrates the difficulty of empirically establishing that policy *directly* benefits/dis-benefits particular groups, the importance of 'spatial nuance' and the necessity of theorisation based upon empirical data rather than just logical assumptions or ideology (McKee *et al*, 2017, p. 60; Soja, 2010).

The Green Belt and Addressing Broader Community Opposition to Greenfield Housebuilding

Moreover, the data showing that opposition to development is shaped by multiscalar fears of change including place attachment, environmental psychology, planning principles and the 'politics of affect' rather than *purely* economic reasons suggests that the house price hypothesis literature on community involvement has privileged the economic too much compared to the emotional and culture (see Chapter 7; Thrift, 2004, p. 57; Harrison and Clifford, 2016). The critical importance of vision and imagination in planning emerged as a key way that planners and campaigners 'frame' the future regarding the Green Belt at multiple spatial scales (Valler and Phelps, 2018, p. 1). The book found that ultimately, these competing visions need to be put to the public in a national and regional Green Belt debate and that the broader planning system needs to better incorporate 'community' knowledge (Bradley, 2018, p. 27). Indeed, public deliberation is probably a more effective way to lead people to consider new solutions to the housing crisis than forced policy solutions and the importance of popular vision and imagination need greater recognition in research as a motivation for campaigners (Inch *et al*, 2020; Sturzaker, 2011).

The Green Belt and the Politics of Planning

The importance of politics in planning emerged particularly strongly regarding the Green Belt. Although the politics of planning has long been recognised in research (e.g. Albrechts, 2020; Cherry, 1982), this study aimed to go further than other studies, especially economic models, by seeking to directly address the 'political' in policy recommendations (Mace, 2018). However, there is a tension between ensuring that planning has democratic legitimacy through politics but, crucially and more philosophically, that it also serves the 'public' interest, especially solving both the housing crisis and climate change, when the system is so politicised as often marked by short-termism, partisanship and lobbying (Inch, 2018; Maidment, 2016; Tait, 2016). Again, this highlights the urgent need for strategic planning, especially of the Green Belt *and* housing numbers (Goode, 2022B).

Planning Policy and Governance

The Pressing Need for Strategic Planning to Address the Housing and Environmental Crises and Other Issues

The proposed policy changes to the Green Belt in this book, including a sustainability purpose underpinned by social and environmental objectives to address the affordability of new homes built on former Green Belt sites and help create a 'greener' policy, have already been outlined in Chapter 6. Consequently, broader structural and governance reforms, centred around the interlocking themes of better systematic integration and more positive planning, are focused on here. The Green Belt's historic success at containing urban areas is acknowledged alongside the effectiveness of the broader planning system in preserving the English countryside in comparison with other countries, but planning is arguably more limited in England when it comes to positively delivering high-quality new development (Sturzaker and Mell, 2016; Raynsford Review, 2018A, 2018B). The book throughout has outlined how many of these issues related to the Green Belt stem from the lack of strategic planning and conceptualised alternative arrangements to address this including underlining the potentiality of 'intermediate' spaces of governance (Chapter 9; Goode, 2022B).

There is a role for the national Government to set up sufficient strategic governance arrangements and to more effectively integrate planning nationally, regionally and locally given the twin imperatives of solving the housing crisis and climate change (Raynsford Review, 2018B; Harris, 2020). The book has made the case for strategic decisions regarding the Green Belt and growth/restraint locations to be *ideally* made strategically through 'intermediate' spaces of deliberation and flexible framework development although it recognises that combined authorities conducting Green Belt reviews are probably the most feasible option in the short-medium term (Goode, 2022B). It remains to be seen if the new Labour Government takes up these recommendations.

Although planning is limited in what it can achieve, there is a case for stronger, more positive and integrated spatial planning along European lines as in the Netherlands and Germany, for example (Goode, 2023; Nadin and Stead, 2008). However, making this 'case' has been challenging given the historical antipathy towards strategic planning of the (until recently) governing Conservative Party, which abolished regional planning in 2010, whilst strategic planning was largely absent from the Levelling Up and Regeneration Act although the Devolution White Paper has the requirement for strategic planning (MHCLG, 2024; DLUHC, 2023; Young, 2023). Indeed, its necessity remains and has become increasingly poignant with the deepening housing crisis and failure of the localism agenda, especially the Duty to Cooperate, to meet the previous Government's 'target' of 300,000 new homes annually (which remains under the Labour Government) (Bradley, 2023; Goode, 2023; McGuinness and Mawson, 2017).

Spatial Scales and the Decision-Making Process

The adversarial, confrontational 'Decide-Announce-Defend' (DAD) nature of planning in many planning systems ideally needs to move towards more proactively and productively engaging people in planning and positively redirecting their energies towards a positive vision of place (Rydin, 2011, p. 95; Raynsford Review, 2018B). Neighbourhood planning in England has been relatively successful in this regard, but the challenge remains of more productively involving people in strategic planning (Cahill, 2004; Wargent and Parker, 2018). Underlying this is the fundamental governance challenge of the appropriate spatial scale at which decisions are made (subsidiarity), strategic decision-making actors and getting governance structures 'right' (Goode, 2022B). Some planning decisions arguably need to be made strategically, such as strategic development sites, housing numbers and the Green Belt, with less scope for *local* politics. Perhaps in 'exchange' for less *local* politics in these higher-level, strategic decisions, there could be greater scope for more community and political engagement with other matters through local plans, such as design or community facilities. Although entailing more governance layers, Table 10.1 develops distinctive, idealised spatial scales of planning decision-making, drawing inspiration from the preceding empirical chapters and the Raynsford Review (2018B, p. 103).

These policy recommendations are necessarily broad because, as the Raynsford Review (2018A) and Planning White Paper (MHCLG, 2020A) alongside Rydin (2011) highlighted, there is value in looking at the *whole* planning system holistically. Indeed, this often distinguishes academic planning research from practitioner studies which usually look at one particular issue or policy (Rydin, 1985, 2013). Conversely, whereas the Raynsford Review (2018A) did not examine policy *per se*, this research has looked at a poignant policy, the Green Belt, and *then* explored the broader interlocking issues related to it in the empirical chapters (the housing crisis (Chapter 6), community opposition (Chapter 7), planning's political nature (Chapter 8) and governance (Chapter 9)).

Planning Practice

The Profession

Chapter 7 highlighted the need for planners to better engage the public in planning, especially to involve a wider range of people, to rebuild people's trust in system systems (Ellis, 2020; Grosvenor, 2019). This is vital to developing a more inclusive, balanced popular 'conversation' on the Green Belt. Another key challenge is better harnessing the public's tacit, experiential knowledge with this book proposing both a national and strategic Green Belt debate. As a minimum, it is important that planners take consultation sincerely as an inherent 'good' although perhaps the issue is more 'who'/'how' the *system* engages people (Upton, 2019; Goode, 2022A). In view of the widespread criticism that it is largely older people with property who 'shout the loudest' in community opposition and the Green Belt

Table 10.1 Idealised governance scales

Spatial scale	Decisions made	Extent of community involvement
National	A national plan setting out key transport priorities, economic objectives and broad locations of development and restraint, such as the Oxford-Cambridge Arc. The overall spatial extent and purpose of the Green Belt could be debated as part of this.	People would be involved in the Green Belt debate. Whilst there would be broad involvement with the national plan, there would be limited *direct* input with it being developed by the central Government.
Strategic (sub-regional or city region)	Setting out strategic transport, land use and economic priorities. A dynamic framework of strategic development locations and Green Belt boundaries, potentially based on recommendations by a Green Belt Commission or Green Growth Board.	The public would be consulted on the overall spatial visions/ options, like new towns, although actual decisions would be made by politicians alongside experts on behalf of residents or by strategic planning advisory boards.
Local	There would be greater public involvement, but the focus would be on the character of development and ensuring supporting facilities rather than large, controversial strategic allocations or housing numbers.	High as with the local plan process now.
Neighbourhood	These would be integrated into local plans and updated at the review stage (Wargent and Parker, 2018). They would, as now, be concerned with the local character of areas and housing, especially its type, and maybe consider *small-scale* Green Belt release.	Very high as with the current process.

debate (Lloyd, 2006, p. 10), planners need to more critically consider *how* they engage the public, including the possibility of better utilising technology as outlined in the Planning White Paper (MHCLG, 2020A). However, technology is still being developed and physical interaction continues to have its value (Wilson *et al*, 2019).

The Profession's Relationship with the Government, Policy, and Academia

It is important for planners and professionals more broadly to consider policy improvement although, given work and time pressures and the seeming lack of Government responsiveness to their views on many issues, such as permitted development rights, resourcing and strategic planning, it is understandable why they give limited consideration to it (Harris, 2020; Hills, 2020). Nonetheless, a significant general challenge in this project was encouraging planners, especially younger ones, to critically reflect and constructively consider what the Green Belt *could* be. Perhaps this was somewhat inevitable for England's most popular and renowned planning policy. Nonetheless, it reflects the broader literature on the 'calculative turn' in planning, its 'fragmentary', increasingly compartmentalised nature and growing proceduralism and 'box-ticking', which has left planners with little 'space' to critically reflect on policy effectiveness and the operational workings of the system overall (McAllister, 2017, p. 122; Parker *et al*, 2018, p. 734; Slade *et al*, 2019, p. 31). Nevertheless, considering policy improvement is still vital for academics and practitioners (Wargent and Parker, 2018) so that proposals are made to the Government and planners and academics speak 'truth unto power' (Tickell, 1995, p. 237).

Finally, many planners interviewed expressed an interest in academic research, but most complained that it was not 'relevant' enough to practice, too 'theoretical' and 'inaccessible' behind the paywalls of academic journals. Of course, theoretical work has its value, and planners may not be interested in or committed to exploring academic work (Dorling and Shaw, 2002; Martin, 2001; Soja, 2010). Additionally, as Bicquelet-Lock (2019) highlighted, practitioner and academic research are distinctive as often having different aims and priorities, but the key is how academic research is *communicated* and *disseminated* to ensure that it is accessible (Kraftl *et al*, 2018). There also needs to be more mutual respect, dialogue and exchange between the academy and practice (Bicquelet-Lock, 2019). Finally, there is a role for professional bodies, such as the RTPI/TCPA, and individual planners to continue to make the case for more powerful and positive planning systems (Kenny, 2019B; Slade *et al*, 2019). This includes planning systems being sufficiently resourced, having the necessary powers related to key issues, such as land assembly, political support to ensure sufficient confidence in and effective functioning of the profession and being able to deliver well-planned new communities alongside preventing inappropriate development.

Housing Crises, Land, Planning Systems and Geography

The Research 'Gap' (See Chapter 1)

This book has filled an important research gap of geographically based, empirical work on the Green Belt and planning systems which both theorise the policy *and* have practical policy relevance to the increasingly pressing 'wicked problem' of the housing crisis (Lund, 2017, p. 36). This affects many countries around the world and is the most significant driver of inequality in England. In contrast to the often aspatial, radical recommendations of the economic school, the book has been cognisant of the Green Belt's institutional history and popular political support in developing realistic recommendations as seeking to move the polarised popular debate forward. It has incorporated theoretical insights, vital for an academic project and drawn from the literature, opposing viewpoints from speaking to a range of stakeholders and consideration of what is politically feasible. The views of *planners* on the policy, which have been missing from many studies on the subject, have been crucially important to this research.

Another key gap that this book has sought to meet is an exploration of the importance of space and geographical variability through a case study of the West Midlands, whereas the Green Belt debate has focused largely on Metropolitan Green Belt whilst first-tier cities, such as Vancouver and San Francisco, have dominated housing debates. Indeed, the book has drawn on a range of multi-scalar sources of evidence through mixed methods compared to the somewhat narrow focus of some academic studies or private sector consultants' reports which largely focus on national statistics and quantitative data (for example, Broadway Malyan, 2015; Hilber and Vermeulen, 2014). Finally, as deriving theoretical insights from the geography/planning literature, the book sits at the intersection of theory, policy and practice which too often have been artificially separated.

Limitations

There are limitations to both the breadth and depth of the book (see Chapter 2). Firstly, analysing the Green Belt using a social justice approach was necessarily broad, and it was challenging to establish causal links, such as the link between the Green Belt and house prices. However, the Green Belt could have been analysed more closely by assessing it directly against the five purposes as would typically be done in a consultant's report. Nonetheless, whilst some of the details and nuances of the policy's effectiveness may have been lost through this study's broad analytical approach, it was vital for the book to be grounded in the very topical and international issue of the housing crisis. Furthermore, using an established, academic analytical approach to social justice made it more related to the academic literature.

Secondly, the findings and theorisation about *everyday* campaigners at the *regional* level were based on extensive engagement with one campaign group (Save Stourbridge Green Belt) and limited engagement with another one (Project Fields). On reflection, there may have been an overfocus on interviewing planners and *professional* campaigners regionally rather than further engagement with everyday campaign groups in the West Midlands. This could potentially have led to planning 'knowledge' being privileged over 'community' knowledge (Bradley, 2018, p. 25), especially with the author being a planner, but given the restricted time frame of data collection, the time-limited nature of campaigners and difficulties of engaging with campaigners, such as Save Stourbridge Green Belt, this would have been much more logistically challenging. Indeed, more qualitative research is needed to focus on the characteristics of a range of campaign groups in different regions/countries and examine whether opposition to Green Belt development varies significantly in character to greenfield development in the UK and in other countries. Finally, in exploring the importance of space, the study could have undertaken more geographically based case studies, especially of other regions such as the North West and North East (Sturzaker, 2010) as well as other countries such as Australia and the USA. This research largely focused on the West Midlands, but other post-industrial regions, which have even more brownfield land and greater urban regeneration challenges, may have provided more spatial nuance on the policy's effectiveness and further elucidated its complex relationship with the housing crisis. However, the methodological reasons for focusing on the 'depth' of a single case study and not exploring a comparative international study were explored in Chapter 2.

Further Research

A wider area of further research could involve comparing the Green Belt with similar policies, such as Urban Growth Boundaries, alongside development management in cities and countries without Green Belts. For example, the Green Belt in England could be compared with the Scottish Green Belt policy as Scotland has a more flexible approach to it and research could probe the potential reasons for this, such as a different popular conception of the countryside and the amount of developed and undeveloped land (explored with Planning Academic (4) in interview) (Lloyd and Peel, 2007). Additionally, comparing the Green Belt in England to the Welsh one could be very instructive, especially for exploring the reasons why Wales only has one Green Belt (Cardiff/Newport) and examining the broader institutional context with the Welsh 2015 Future Generations Act making statutory commitments on the sustainable development goals (Carter and MacKillop, 2024).

More broadly, comparing the English policy to other green belts internationally, both present and former Green Belts, such as those in Christchurch, Tokyo, Sydney and Seoul would be useful to evaluate their effectiveness, chart their spatial extent and explore their governance arrangements and the reason for their demise compared to the longevity and success of the Green Belt in England. The effectiveness of the Green Belt as a policy needs to be compared with similar policies, such as Urban Growth Boundaries in Portland and Melbourne, whilst the Green Belt could be compared with other growth management policies, such as green wedges in Leicester or the 'finger plan' in Copenhagen, for example. Cities with a Green Belt could be compared to demographically and economically similar cities without Green Belts, such as Hull/Liverpool, Exeter/York and Leicester/Derby, to research how different LPAs manage urban growth and further evaluate the Green Belt's effectiveness, especially its effects upon house prices etc.

Finally, the Green Belt in England could be compared to growth management in other countries, such as Ireland, whose planning system has a similar origin to Britain's but whose attitude and approach to development are different (Fox-Rogers and Murphy, 2014, 2016). Likewise, it would be useful to compare the conflict over the Green Belt in England with the conflict over greenfield housing development in similar contexts internationally such as the Groene Hart ('Green Heart') in the Netherlands and Urban Growth Boundaries in places like Portland and indeed the failure of Green Belts in Southeast Asia and New Zealand (Amati, 2007; Boyle and Mohamed, 2007; Altes, 2017).

Research is also needed on the failures of Green Belts internationally. Amati (2007), for example, related the failure of Green Belts that were introduced in Southeast Asia to being based upon a Western dualism of the separation of rural and urban which contrasted with the 'mixed' landscapes in Southeast Asian countries where rice paddy fields were interspersed with urban development. Additionally, there could be a comparison with countries with looser planning systems, such as the USA and Australia, and those with stronger planning systems, like Germany, the Netherlands or China.

A fruitful area of wider research could be the land market and landowner attitudes/behaviour in different countries (Adams and Watkins, 2014). For example, why has the Green Belt been so long-standing in England compared to other countries, such as Japan and Australia, where landowner pressure forced Governments to abandon attempts at green belts (Amati, 2008; Sorensen, 2002)? The notion of landowner beneficence, benevolence and conservatism in England, briefly referenced in this book in Chapter 3 (Amati and Yokohari, 2003, 2006, 2007), could be a fascinating area of further research. Additionally, the wide difference in landowner attitudes towards developing their land, especially in the Green Belt, could be researched temporally and geographically. For example, why is it that some landowners aggressively lobby and promote their land for development, perhaps through a land promoter (Lichfields, 2018), whereas other landowners are more conservative and satisfied for their land to remain in agricultural use? Specifically, the existence of 'options' between landowners and housebuilders has been highlighted in Chapter 6 (Monbiot *et al*, 2019, p. 16). However, given the lack of knowledge about them, research exploring their nature and characteristics, geographical extent and landowner attitudes towards them could be a fruitful area of future research, especially in comparison with other land markets internationally.

Adams and Hutchison (2000) have conducted valuable research into landowner behaviour on brownfield sites, but there is a need for more research on Green Belt and greenfield sites both in England and internationally. Indeed, this is a prescient issue given the widespread public interest in landownership in England and the cost of land and land value captured following the popular publication *Who Owns Britain?* (Shrubsole, 2019) and MHCLG's (2020B) consultation and call for evidence on *transparency, competition and land control*. In particular, with the high cost of land in England (Cheshire, 2014, 2024), the question of how far house prices are more affordable in countries with low(er) land prices, such as the USA and the Netherlands, and the broader relationship between land/house prices and economic growth are extremely important for policy and planning (Adams and Watkins, 2014).

The housebuilding industry's structure and how it affects housing output and prices is another pressing issue in England. Academic and practitioner research (i.e. Adams and Tiesdell, 2013; Barlow and King, 1992; Letwin, 2018), has helpfully explored the extent to which the market is dominated by volume housebuilders, highlighted some impacts upon housing delivery and made the case for more SME (Small and Medium-Sized Enterprises) builders. However, more research is needed on the impact of the industry's structure upon house prices and output, perhaps through cross-comparison with other European countries that support more self and custom build such as the Netherlands and Germany (for example, Barlow and King's (1992) comparative study of Sweden, France and England). There is also the issue of the dominant tenure and type of housing built in Britain compared to other countries, which could be researched, alongside other issues related to the Green Belt, but this conclusion has highlighted the most pressing topics for further research.

Final Remarks and Reflections

To research Green Belts is to research space as most of the wide-ranging issues connected with them – housing crises, community opposition, the political nature and governance of planning systems and green infrastructure – are also inherently.

Firstly, housing markets are marked by spatial heterogeneity which characterise their interactions with Green Belts, Urban Growth Boundaries and other planning policies and means that disentangling the specific impacts of planning policies on house prices from other factors is challenging. This book has demonstrated the importance of space and geography as the key components of the issues related to the Green Belt *and* at the heart of policy recommendations.

Secondly, the book has argued that community opposition to housing development often acquires specific geographical characteristics which contribute to some extent towards the conceptual binary of 'community' knowledge being viewed as more emotional, experiential juxtaposed to technical, objective or even 'scientific' planning knowledge (Bradley, 2018, p. 24). Thirdly, based on the geographically based case study of power in planning systems in this book, the conceptualisation of the gap between *attempted* and *effective* exercise of power was developed. Moreover, the book has examined how many of these issues related to the Green Belt stem from the lack of geographically based strategic planning and conceptualised alternative arrangements to address this including underlining the potentiality of 'intermediate' spaces of governance. Fourthly, in relation to green infrastructure, the book has demonstrated the outdatedness and challenge of responding to the modernist dualism and separation of rural and urban whilst examining the potentiality of a more environmentally focused, dynamic Green Belt policy.

Green Belts are therefore a vitally important lens through which space more broadly can be theorised whilst space, spatial configurations and social relations associated with it are themselves shaped by the policy. More broadly, the Green Belt is an extremely useful object of study because, as this book has found, it yields valuable wider insights into many areas, including society, economy and culture, so is a rich seam for conceptualisation and theorisation both in the UK and internationally. As probably the most long-standing and popular planning policy in England, it has important implications for planning theory, policy and practice.

References

Aalbers, M. (2019) 'Introduction to The Forum: From Third to Fifth-Wave Gentrification', *Tijdschrift voor Economische en Sociale Geografie*, 110(1), pp. 1–11.

Abbott, L. (2013) *Political Barriers to Housebuilding in Britain* (Manchester: Industrial Systems Research).

Abercrombie, P. (1944) *Greater London Plan 1944* (London: HMSO).

Adams, D. and Hutchison, N. (2000) 'The urban task force report: Reviewing land ownership constraints to brownfield redevelopment', *Regional Studies*, 34(2), pp. 777–782.

Adams, D. and Leishman, C. (2008) *CLG Housing Markets and Planning Analysis Expert Panel Factors Affecting Housing Build-Out Rates* (Glasgow: University of Glasgow).

Adams, D. and Tiesdell, S. (2013) *Shaping Places: Urban Planning, Design and Development* (Abingdon: Routledge).

Adams, D. and Watkins, C. (2014) *The Value of Planning* (London: RTPI).

AECOM (2015) *Big Bold Global Connected London 2065* (London: AECOM).

Airey, J. and Doughty, C. (2020) *Rethinking the Planning System for the 21st Century* (London: Policy Exchange).

Albrechts, L. (2020) *Planners in Politics: Do They Make a Difference?* (Cheltenham: Elgar).

Alexander, E. (2010) 'Planning, Policy and the Public Interest: Planning Regimes and Planners' Ethics and Practices', *International Planning Studies*, 15(2), pp. 143–162.

Allen, J. (2003) *Lost Geographies of Power* (Oxford: Blackwell Publishing).

Allen, J. (2011) 'Topological Twists: Power's Shifting Geographies', *Dialogues in Human Geography*, 1(3), pp. 283–298.

Allen, J. and Cochrane, A. (2010) 'Assemblages of State Power: Topological Shifts in the Organisation of Government and Politics', *Antipode*, 42(5), pp. 1071–1089.

Allmendinger, P. (2009) *Planning Theory* (Basingstoke: Palgrave).

Allmendinger, P. and Haughton, G. (2011) 'Challenging Localism', *Town and Country Planning*, 80(7/8), pp. 314–317.

Allmendinger, P. and Haughton, G. (2012) 'Post-Political Spatial Planning in England: A Crisis of Consensus?', *Transactions of the Institute of British Geographers*, 37, pp. 89–103.

Altes, W. (2017) 'Rules Versus Ideas in Landscape Protection: Is a Green Heart Attack Imminent?', *International Planning Studies*, 23(1), pp. 1–15.

Altshuler, A. (1966) *City Planning Process: A Political Analysis* (New York: Cornell University Press).

Amati, M. (2007) 'From a Blanket to a Patchwork: The Practicalities of Reforming the London Green Belt', *Journal of Environmental Planning and Management*, 50(5), pp. 579–594.

Amati, M. (ed.) (2008) *Urban Green Belts in the Twenty-First Century* (London: Ashgate).

Amati, M. and Taylor, L. (2010) 'From Green Belts to Green Infrastructure', *Planning Practice and Research*, 25(2), pp. 143–155.

Amati, M. and Yokohari, M. (2003) 'The Actions of Landowners, Government and Planners in Establishing the London Green Belt of the 1930s', *Journal of the Japanese Institute of Landscape Architecture*, 67(5), pp. 433–438.

Amati, M. and Yokohari, M. (2004) 'The Actions of Landowner, Government and Planners in Establishing the London Green Belt of the 1930s', *Planning History*, 26(1–2), pp. 433–438.

Amati, M. and Yokohari, M. (2006) 'Temporal Changes and Local Variations in the Functions of London's Green Belt', *Landscape and Urban Planning*, 75(1–2), pp. 125–142.

Amati, M. and Yokohari, M. (2007) 'The Establishment of the London Greenbelt: Reaching Consensus Over Purchasing Land', *Journal of Planning History*, 6(4), pp. 311–337.

Anton, C. and Lawrence, C. (2014) 'Home Is Where the Heart Is: The Effect of Place of Residence on Place Attachment and Community Participation', *Journal of Environmental Psychology*, 40, pp. 451–461.

Archer, T. and Cole, I. (2014) 'Still Not Plannable?', *People, Place and Policy Online*, 8(2), pp. 97–112.

Bailey, E., Devine-Wright, P. and Batel, S. (2016) 'Using a Narrative Approach to Understand Place Attachments and Responses to Power Line Proposals: The Importance of Life-Place Trajectories', *Journal of Environmental Psychology*, 48, pp. 200–211.

Baker, J. (2018) *A Point of View on Green Belts* (Taunton: RTPI South West).

Balen, M. (2006) *Land Economy: How a Rethink of How Our Will Benefit Britain* (London: Adam Smith Institute).

Ball, M. (2004) 'Co-Operation With the Community in Property-Led Urban Regeneration', *Journal of Property Research*, 21(2), pp. 119–142.

Ball, M., Cigdem, M., Taylor, E. and Wood, G. (2014) 'Urban Growth Boundaries and Their Impact on Land Prices', *Environment and Planning A*, 46(12), pp. 3010–3026.

Barber, A. and Hall, S. (2008) 'Birmingham: Whose Urban Renaissance? Regeneration as a Response to Economic Restructuring', *Policy Studies*, 29(3), pp. 281–292.

Barker, K. (2004) *Review of Housing Supply – Delivering Stability: Securing Our Future Housing Needs, Interim Report* (London: TSO).

Barker, K. (2006) *Barker Review of Land-Use Planning, Final Report – Recommendations* (London: HM Treasury).

Barlow, J. and King, A. (1992) 'The State, the Market and Competitive Strategy: The Housebuilding Industry in the United Kingdom, France, and Sweden', *Environment and Planning A*, 24, pp. 381–400.

Barry, J., Horst, M., Inch, A., Legacy, C., Rishi, S., Rivero, J., Taufen, A., Zanotto, J. and Zitcer, A. (2018) 'Unsettling Planning Theory', *Planning Theory*, 17(3), pp. 418–438.

Best, M. (2019) 'The Lost Decade, Best Laid Plans', https://mikesbestlaidplans.wordpress. com/ (Accessed: 8/1/2024).

Bicquelet-Lock, A. (2019) 'Clearing a Path Between Planners, Planning Policy and Academics – Challenges and Opportunities', *Town and Country Planning*, 88(1), pp. 15–20.

Birmingham City Council (2017) 'Birmingham Plan 2031: Birmingham Development Plan', https://www.birmingham.gov.uk/downloads/file/5433/adopted_birmingham_ development_plan_2031 (Accessed: 10/4/2024).

Blake, R. and Golland, A. (2003) *Housing Development* (London: Routledge).

Blyth, R. (2017) 'My Back Yard', RTPI, http://www.rtpi.org.uk/briefing-room/rtpi-blog/my-back-yard/ (Accessed: 10/4/2024).

Boddy, M. and Hickman, H. (2018) '"Between a Rock and a Hard Place": Planning Reform, Localism and the Role of the Planning Inspectorate in England', *Planning Theory and Practice*, 19(2), pp. 198–217.

Boddy, M. and Hickman, H. (2020) '"If Independence Goes, the Planning System Goes": New Political Governance and the English Planning Inspectorate', *Town Planning Review*, 91(1), pp. 21–45.

Bourdieu, P. (2001) *Masculine Domination* (Second ed., Stanford, CA: Stanford University Press).

Boyle, R. and Mohamed, R. (2007) 'State Growth Management, Smart Growth and Urban Containment: A Review of the US and a Study of the Heartland', *Journal of Environmental Planning and Management*, 50(5), pp. 677–697.

Bradley, Q. (2018) 'Neighbourhood Planning and the Production of Spatial Knowledge', *Town Planning Review*, 89(1), pp. 23–42.

Bradley, Q. (2021) 'The Financialisation of Housing Land Supply in England', *Urban Studies*, 58(2), pp. 389–404.

Bradley, Q. (2023) *Property, Planning and Protest: The Contentious Politics of Housing Supply* (Abingdon: Routledge).

Bradley, Q. (2019A) 'Combined Authorities and Material Participation: The Capacity of Green Belt to Engage Political Publics in England', *Local Economy*, 34(2), pp. 181–195.

Bradley, Q. (2019B) 'Public Support for Green Belt: Common Rights in Countryside Access and Recreation', *Journal of Environmental Policy and Planning*, 21(6), pp. 692–701.

Bramley, G. (1993) 'Land-Use Planning and the Housing Market in Britain: The Impact on Housebuilding and House Prices', *Environment and Planning A*, 25(7), pp. 1021–1057.

Bramley, G. (2019) *Housing Supply Requirements Across Great Britain: For Low-Income Households and Homeless People* (Edinburgh: Crisis and National Housing Federation).

Bramley, G., Morgan, J., Ballantyne Way, S., Cousins, L. and Houston, D. (2017) 'The Deliverability and Affordability of Housing in the South West of England', RTPI, https://www.rtpi.org.uk/media/1949/deliverability-and-affordability-of-housing-in-the-south-west-of-england-full-report.pdf (Accessed: 16/07/2023).

Brand, R. and Gaffikin, F. (2007) 'Collaborative Planning in an Uncollaborative World', *Planning Theory*, 6(3), pp. 282–313.

Branson, A. (2019) 'How Joint Strategic Plans Are Progressing Across England, and the Obstacles that They Face', *Planning Resource*, https://www.planningresource.co.uk/article/1591341/joint-strategic-plans-progressing-across-england-obstacles-face (Accessed: 1/08/2023).

Brassington, J. (2019) 'Boris Johnson Backs Campaign to Save Wolverhampton Green Belt', *Express and Star*, https://www.expressandstar.com/news/politics/2019/09/20/boris-johnson-backs-campaign-to-save-wolverhampton-greenbelt/ (Accessed: 30/01/2023).

Breheny, M. (1997) 'Urban Compaction: Feasible and Acceptable?', *Cities*, 97(4), pp. 209–217.

Broadway Malyan (2015) '50 Shades of Green Belt Report', https://issuu.com/broadwaymalyan/docs/50_shades_of_green_belt_report (Accessed: 10/07/2024).

Brock, A. (2018) 'Guildford Local Plan Defended by Council at Public Examination Amid Criticism from Green Belt Campaigners', *Surrey Live*, https://www.getsurrey.co.uk/news/surrey-news/guildford-local-plan-defended-council-14756578 (Accessed: 13/09/2023).

Bromley (2019) 'Draft London Plan EiP 2019', https://www.london.gov.uk/sites/default/files/m10_lb_bromley_2593.pdf (Accessed: 8/1/2022).

Brownill, S. and Inch, A. (2019) 'Framing People and Planning: 50 Years of Debate', *Built Environment*, 45(1), pp. 7–25.

Bruegmann, R. (2005) *Sprawl: A Compact History* (Chicago, IL: University of Chicago).

Bryman, A. (2012) *Social Research Methods* (Oxford: Oxford University Press).

Buckle, C., Simmie, H. and Formston, D. (2024) 'English Housing Supply Update Q4 2023', *Savills*, https://www.savills.co.uk/research_articles/229130/357082-0#summary (Accessed: 03/04/2024).

Budnitz, H., Tranos, E. and Chapman, L. (2020) 'Telecommuting and Other Trips: An English Case Study', *Journal of Transport Geography*, 85, pp. 1–9.

Bunce, M. (1994) *The Countryside Ideal: Anglo-American Images of Landscape* (London: Routledge).

Cahill, C. (2004) 'Defying Gravity? Raising Consciousness Through Collective Research', *Children's Geographies*, 2(2), pp. 273–286.

Cameron, J. *et al.* (2005) 'Focusing on the Focus Group', in Hay I. (ed.) *Qualitative Research and Human Geography* (Oxford: Oxford University Press), pp. 83–101.

Campbell, H., Tait, M. and Watkins, C. (2014) 'Is There Space for Better Planning in a Neoliberal World? Implications for Planning Practice and Theory', *Journal of Planning Education and Research*, 34(1), pp. 45–59.

Carlton, I. (2007) *Histories of Transit-Orientated Development: Perspectives on the Development of the TOD Concept* (Berkeley: University of California).

Carpenter, J. (2016) 'What the Call-in of Birmingham's Local Plan Means for Planmakers', *Planning Resource,* http://www.planningresource.co.uk/article/1398193/call-in-birminghams-local-plan-means-plan-makers (Accessed: 03/04/2024).

Carter, I. and MacKillop (2024) 'Can We Promote Plural Local Pathways to Sustainable Development? Insights from the Implementation of Wales's Future Generations Act', *Journal of Environmental Policy and Planning*, 25(5), pp. 554–569.

Castells, M. (1983) *The City and the Grassroots* (Berkeley and Los Angeles: University of California Press).

Castree, M. *et al.* (2008) 'The Limits to Capital (1982): David Harvey', in Hubbard P., Kitchin P. and Valentine G. (eds.) *Key Texts in Human Geography* (London: SAGE), pp. 61–70.

Catney, P. and Henneberry, J. (2019) 'Change in the Political Economy of Land Value Capture in England', *Town Planning Review*, 90(4), pp. 339–358.

Cherry, G. (1982) *Politics of Town Planning* (London: Longman).

Cherry, G. (1996) *Town Planning in Britain Since 1900* (Oxford: Blackwell Publishers).

Cheshire, P. (2009) 'Urban Containment, Housing Affordability and Price Stability – Irreconcilable Goals', *SERC Policy Paper*, 4, pp. 1–20.

Cheshire, P. (2013) 'Greenbelt Myth Is the Driving Force Behind the Housing Crisis', *LSE Blogs*, http://blogs.lse.ac.uk/politicsandpolicy/greenbelt-myth-is-the-driving-force-behind-housing-crisis/ (Accessed: 31/7/2018).

Cheshire, P. (2014) 'Turning Houses into Gold: The Failure of British Planning', *CentrePiece Spring*, pp. 14–18.

Cheshire, P. *et al.* (2014) 'Decent Housing or Rigid Greenbelts?', in Manns J. (ed.) *Kaleidoscope City: Reflections on Planning and London* (London: Colliers International and RTPI), pp. 153–161.

Cheshire, P. (2019) 'The Costs of Containment: Or the Need to Plan for Urban Growth', *CESifo Forum*, 20(3), pp. 10–14.

Cheshire, P. (2024) 'Pushing Water Uphill: Containment Policies Doomed to Fail', *Town Planning Review, Ahead of Print*, pp. 1–15.

Cheshire, P. and Buyuklieva, B. (2019) 'Homes on the Right Tracks: Greening the Green Belt to Solve the Housing Crisis', *Centre for Cities,* www.centreforcities.org/about (Accessed: 4/1/2024).

Chettiparamb, A. (2016) 'Articulating "Public Interest" Through Complexity Theory', *Environment and Planning C: Government and Policy*, 34(7), pp. 1284–1305.

Christophers, B. (2018) 'Intergenerational Inequality? Labour, Capital, and Housing Through the Ages', *Antipode*, 50(1), pp. 101–121.

Christophers, B. (2019) 'A Tale of Two Inequalities: Housing-Wealth Inequality and Tenure Inequality', *Environment and Planning A*, 53(3), pp. 573–594.

Chu, E., Anguelovski, I. and Roberts, D. (2017) 'Climate Adaptation as Strategic Urbanism: Assessing Opportunities and Uncertainties for Equity and Inclusive Development in Cities', *Cities*, 60, pp. 378–387.

Cidell, J. (2005) 'Reviewed Work(s) Lost Geographies of Power and Geographies of Power: Placing Scale', *Annals of the Association of American Geographers*, 95(1), pp. 230–232.

Clarke, E., Nohrova, N. and Thomas, E. (2014) *Delivering Change: Building Homes Where We Need Them* (London: Centre for Cities).

Clews, R. (2016) 'Report on the Examination of the Birmingham Development Plan "Birmingham Plan 2031"', *Planning Inspectorate*, file:///C:/Users/goodec/Downloads/BDP_Inspectors_Report.pdf (Accessed: 4/1/2024).

Coelho, M., Ratnoo, V. and Dellepiane, S. (2014) *Housing That Works for All: The Political Economy of Housing in England* (London: Institute for Government).

Colomb, C. and Tomaney, J. (2016) 'Territorial Politics, Devolution and Spatial Planning in the UK: Results, Prospects, Lessons', *Planning Practice and Research*, 31(1), pp. 1–22.

Community Research (2019) 'Citizens' Jury – Future of London's Green Belt', https://www.londonfirst.co.uk/sites/default/files/documents/2019-09/CitizensJury.pdf (Accessed: 21/02/2022).

ComRes (2018) 'Centre for Policy Studies – Housing Poll September 2018', https://cps.org.uk/wp-content/uploads/2023/01/CPS_THE_CASE_FOR_HOUSEBUILDING2.pdf (Accessed: 11/04/2024).

Connell, D. (2009) 'Planning and Its Orientation to the Future', *International Planning Studies*, 14(1), pp. 85–98.

Conservative Party (2019) 'Get Brexit Done: Unleash Britain's Potential', https://www.conservatives.com/our-plan (Accessed: 08/04/2024).

Conti, J. and O'Neil, M. (2007) 'Studying Power: Qualitative Methods and the Global Elite', *Qualitative Research*, 7(1), pp. 63–82.

Corlett, A. and Judge, L. (2017) *Home Affront: Housing Across the Generations* (London: Resolution Foundation).

County Councils Network (2020) 'Evaluating the Importance of Scale in Proposals for Local Government Reorganisation', https://www.countycouncilsnetwork.org.uk/wp-content/uploads/PwC-Evaluating-the-importance-of-scale-in-proposals-for-local-go.vernment-reorganisation.pdf (Accessed: 4/1/2024).

Cowell, R., Bristow, G. and Munday, M. (2011) 'Acceptance, Acceptability and Environmental Justice: The Role of Community Benefits in Wind Energy Development', *Journal of Environmental Planning and Management*, 54(4), pp. 539–557.

Cox, W. (2002) *Property, Prosperity and Poverty: Trends and Choices in Land Use Policy.* London.

CPRE (2005) *Green Belt: 50 Years On* (London: CPRE).

CPRE (2015) '60th Anniversary Poll Shows Clear Support for Green Belt', CPRE, https://www.cpre.org.uk/media-centre/latest-news-releases/item/4033-60th-anniversary-poll-shows-clear-support-for-green-belt (Accessed: 25/01/2023).

CPRE (2018) 'The State of the Green Belt', CPRE, https://www.cpre.org.uk/wp-content/uploads/2019/11/State_of_the_Green_Belt_2018.pdf (Accessed: 4/1/2024).

CPRE and Natural England (2010) *Green Belts: A Greener Future A Joint Report by CPRE and Natural England Green Belts in England* (London: CPRE).

Crook, A. and Whitehead, C. (2019) 'Capturing Development Value, Principles and Practice: Why Is It so Difficult?', *Town Planning Review*, 90(4), pp. 359–381.

Crook, T. (2015) 'Review Policy to Allow Planned Release of Land', *Financial Times*, http://www.ft.com/cms/s/0/966770b8-b074-11e4-a2cc-00144feab7de.html#axzz4BkHENXXf (Accessed: 6/02/2024).

Crossman, R. (1975) *The Diaries of a Cabinet Minister. Vol 1, Minister of Housing, 1964–66* (London: Hamilton).

Cullingworth, B., Nadin, T., Davoudi, S., Pendlebury, J., Vigar, G. and Webb, D. (2015) *Town and Country Planning in the UK.* 15th edn. (London: Routledge).

Curley, R. (2019) 'Thousands of Homes to be Built in Guildford by 2034 as £3.4m Local Plan Approved', *Surrey Live*, https://www.getsurrey.co.uk/news/surrey-news/guildford-local-plan-approved-councillors-16184309 (Accessed: 14/08/2020).

Daniels, T. (2010) 'The Use of Green Belts to Control Sprawl in the United States', *Planning Practice and Research*, 25(2), pp. 255–271.

Davis, J. and Huennekens, J. (2022) 'YIMBY Divided: A Qualitative Content Analysis of YIMBY Subreddit Data', *Journal of Urban Affairs*, 46(9), pp. 1810–1836.

Davison, G., Han, H. and Liu, E. (2017) 'The Impacts of Affordable Housing Development on Host Neighbourhoods: Two Australian Case Studies', *Journal of Housing and the Built Environment*, 32(4), pp. 733–753.

Davison, G., Legacy, C., Liu, E. and Darcy, M. (2016) 'The Factors Driving the Escalation of Community Opposition to Affordable Housing Development', *Urban Policy and Research*, 34(4), pp. 386–400.

Davoudi, S. and Sturzaker, J. (2017) 'Urban Form, Policy Packaging and Sustainable Urban Metabolism', *Resources, Conservation and Recycling*, 120, pp. 55–64.

Dawkins, C.J. and Nelson, A.C. (2002) 'Urban Containment Policies and Housing Prices: An International Comparison With Implications for Future Research', *Land Use Policy*, 19(1), pp. 1–12.

DCLG (2012) *National Planning Policy Framework* (London: Department for Communities and Local Government).

DCLG (2017) 'Fixing Our Broken Housing Market', Department for Communities and Local Government, https://www.gov.uk/government/uploads/system/uploads/attachment_data/file/590463/Fixing_our_broken_housing_market_-_accessible_version.pdf (Accessed: 7/1/2023).

Dear, M. and Scott, A. (eds.) (1981) *Urbanisation and Urban Planning in Capitalist Societies* (London: Methuen).

Dehring, C., Depken, C. and Ward, M. (2008) 'A Direct Test of the Homevoter Hypothesis', *Journal of Urban Economics*, 64(1), pp. 155–170.

Dembski, S., Sykes, O., Couch, C., Desjardins, X., Evers, D., Osterhage, F., Siedentop, S. and Zimmermann, K. (2021) 'Reurbanisation and Suburbia in Northwest Europe', *Progress in Planning*, 150, pp. 1–48.

Department of the Environment (1995) *Planning Policy Guidance 2: Green Belts* (London: Department of the Environment).

Derbyshire, B. (2015) *Building Greater London: An End to the Capital's Crisis of Affordability* (London: The London Society).

DeVerteuil, G. (2013) 'Where Has NIMBY Gone in Urban Social Geography?', *Social and Cultural Geography*, 14(6), pp. 599–603.

Devine-Wright, P. (2009) 'Rethinking NIMBYism: The Role of Place Attachment and Place Identity in Explaining Place-Protective Action', *Journal of Community and Applied Social Psychology*, 19, pp. 426–441.

DLUHC (2023) 'National Planning Policy Framework', https://webarchive.national-archives.gov.uk/ukgwa/20231228093504/https://www.gov.uk/government/publications/national-planning-policy-framework--2 (Accessed: 02/04/2024).

Dobson, M. and Parker, G. (2024) 'The Temporal Governance of Planning in England: Planning Reform, Uchronia and 'Proper Time'', *Planning Theory*, 0(0), pp. 1–22.Author: URL is not valid, please check and correct

Dockerill, B. and Sturzaker, J. (2019) 'Green Belts and Urban Containment: The Merseyside Experience', *Planning Perspectives*, 35(4), pp. 1–26.

Donnelly, M. (2019) 'Authorities can Ignore Latest Household Projections in Assessing Housing Need, Confirms MHCLG', *Planning Resource*, https://www.planningresource.co.uk/article/1526169/authorities-ignore-latest-household-projections-assessing-housing-need-confirms-mhclg (Accessed: 9/11/2022).

Dorling, D. (2015) *All That Is Solid* (London: Penguin).

Dorling, D. and Shaw, M. (2002) 'Geographies of the Agenda: Public Policy, the Discipline and Its (re)'turns', *Progress in Human Geography*, 26(5), pp. 629–646.

Dowding, K. (2006) 'Three-Dimensional Power: A Discussion of Steven Lukes' Power: A Radical View', *Political Studies Review*, 4(2), pp. 136–145.

Dudley Metropolitan Borough Council, Sandwell Metropolitan Borough Council, Walsall Council and Wolverhampton City Council (2011) 'Black Country Core Strategy', https://blackcountryplan.dudley.gov.uk/t1/p2/ (Accessed: 4/1/2024).

Dudley Metropolitan Borough Council, Sandwell Metropolitan Borough Council, Walsall Council and Wolverhampton City Council (2019) 'Black Country Urban Capacity Review December 2019', https://blackcountryplan.dudley.gov.uk/media/13808/bcp-summary-urban-capacity.pdf (Accessed: 4/1/2024).

Dühr, S. (2018) 'A Europe of "Petites Europes": An Evolutionary Perspective on Transnational Cooperation on Spatial Planning', *Planning Perspectives*, 33(4), pp. 543–569.

Edgar, L. (2015) 'Pickles Rejects Green Belt Business Park', *The Planner*, https://www.theplanner.co.uk/news/pickles-rejects-greenbelt-business-park (Accessed: 12/07/2017).

Edwards, M. (2015) *Prospects for Land, Rent and Housing in UK Cities* (London: Government Office for Science).

Edwards, M. (2016A) 'The Housing Crisis and London', *City*, 20(2), pp. 222–237.

Edwards, M. (2016B) 'Former Wisley Airfield Site Plan for More than 2,000 Homes Unanimously Rejected', *Surrey Live*, http://www.getsurrey.co.uk/news/surrey-news/former-wisley-airfield-site-plan-11150488 (Accessed: 27/06/2023).

Elkes, N. (2016) 'Sutton Coldfield Green Belt Plans for 6,000 Homes put on Hold', *Birmingham Post*, http://www.birminghampost.co.uk/news/regional-affairs/sutton-coldfield-green-belt-plans-11393614 (Accessed: 31/03/2022).

Ellis, H. (2020) 'The Planning White Paper and that Morning After Feeling', *Town and Country Planning*, https://www.tcpa.org.uk/planning-white-paper-morning-after-feeling/ (Accessed: 4/1/2024).

Ellis, H. and Henderson, K. (2014) *Rebuilding Britain: Planning for a Better Future* (London: Town and Country Planning Association).

Elson, M. (1986) *Green Belts: Conflict Mediation in the Urban Fringe* (London: Meinemann).

Elson, M. (2002) 'Modernising Green Belts—Some Recent Contributions', *Town and Country Planning*, 71(10), pp. 266–270

Elson, M., Walker, S. and Macdonald, R. (1994) *The Effectiveness of Green Belts* (London: HMSO Department of the Environment).

Evans, A. (1996) 'The Impact of Land Use Planning and Tax Subsidies on the Supply and Price of Housing in Britain: A Comment', *Urban Studies*, 33(3), pp. 581–585.

Fainstein, S. (2010) *The Just City* (New York: Cornell University Press).

Fainstein, S. (2014) 'The Just City', *International Journal of Urban Sciences*, 18(1), pp. 1–18.

Fairclough, N. (2001) *Language and Power* (Longman: Harlow).

Faludi, A. (1973) *A Reader in Planning Theory* (Oxford: Pergamon Press).

Federation of Master Builders (2018) 'Co-Living and Micro-Homes Most Popular Solutions to Housing Crisis', https://www.fmb.org.uk/about-the-fmb/newsroom/co-living-and-micro-homes-most-popular-solutions-to-housing-crisis/ (Accessed: 4/1/2024).

Ferm, J. and Raco, M. (2020) 'Viability Planning, Value Capture and the Geographies of Market-Led Planning Reform in England', *Planning Theory and Practice*, 21(2), pp. 218–235.

Fernandez, R., Hofman, A. and Aalbers, M. (2016) 'London and New York as a Safe Deposit Box for the Transnational Wealth Elite', *Environment and Planning A*, 48(12), pp. 2443–2461.

Field, A. (2013) *Discovering Statistics Using SPSS*. 4th edn. (London: Sage).

Fischel, W. (2001A) *The Homevoter Hypothesis* (Cambridge: Harvard University Press).

Fischel, W. (2001B) 'Why Are There NIMBYs?', *Land Economics*, 77(1), pp. 144–152.

Flemming, P. (2019) 'Local Plan Hearings Cancelled', Sevenoaks District Council, https://www.sevenoaks.gov.uk/news/article/119/council_concerned_after_local_plan_hearings_are_cancelled (Accessed: 4/1/2024).

Flyvbjerg, B. (1998) *Rationality and Power* (Chicago: University of Chicago Press).

Flyvbjerg, B. (2004) 'Phronetic Planning Research: Theoretical and Methodological Reflection', *Planning Theory and Practice*, 5(3), pp. 283–306.

Foglesong, R. (1986) *Planning the Capitalist City: The Colonial Era to the 1920s* (Princeton, NJ: Princeton University Press).

Foresight Report (2010) *Land Use Futures Project: Final Project Report* (London: Government Office for Science).

Forshaw, J. and Abercrombie, P. (1943) *County of London Plan 1943* (London: HMSO).

Foucault, M. (1977) *Discipline and Punish* (New York: Vintage Books).

Foucault, M. *et al.* (1991) 'Questions of Method', in Burchell G., Gordon C. and Miller P. (eds.) *The Foucault Effect: Studies in Governmentality* (London: Harvester Wheatsheaf), pp. 87–104.

Fox-Rogers, L. and Murphy, E. (2014) 'Informal Strategies of Power in the Local Planning System', *Planning Theory*, 13(3), pp. 244–268.

Fox-Rogers, L. and Murphy, E. (2016) 'Self-Perceptions of the Role of the Planner', *Environment and Planning B: Planning and Design*, 43(1), pp. 74–92.

Gallent, N. (2019) *Whose Housing Crisis? Assets and Homes in a Changing Economy* (Bristol: Policy Press).

Gallent, N., Andersson, J. and Bianconi, M. (2006) 'Planning on the Edge: England's Rural – Urban Fringe and the Spatial-Planning Agenda', *Environment and Behaviour B: Planning and Design*, 33(3), pp. 457–476.

Gallent, N., De Magalhaes, C. and Freire Trigo, S. (2021) 'Is Zoning the Solution to the UK Housing Crisis?', *Planning Practice and Research*, 26(1), pp. 1–19.

Gallent, N., Hamiduddin, I., Stirling, P. and Kelsey, J. (2019) 'Prioritising Local Housing Needs Through Land-Use Planning in Rural Areas: Political Theatre or Amenity Protection?', *Journal of Rural Studies*, 66, pp. 11–20.

Gallent, N. and Shaw, D. (2007) 'Spatial Planning, Area Action Plans and the Rural-Urban Fringe', *Journal of Environmental Planning and Management*, 50(5), pp. 617–638.

Gant, R., Robinson, G. and Fazal, S. (2011) 'Land-Use Change in the "edgelands": Policies and Pressures in London's Rural-Urban Fringe', *Land Use Policy*, 28(1), pp. 266–279.

Geoghegan, J. (2018) 'Did Raab Make Immigration and Housing Comments for Political Reasons', *Planning Resource*, https://www.planningresource.co.uk/article/1462445/raab-immigration-housing-comments-political-reasons-john-geoghegan (Accessed: 7/11/2022).

Gibson, T. (2005) 'NIMBY and the Civic Good', *City and Community*, 4(4), pp. 381–401.

Gilmore, G. (2014) *Building Momentum: Housebuilding Report 2014* (London: Knight Frank).

Gilmore, G. (2016) *Gathering Momentum: Housebuilding Report 2016* (London: Knight Frank).

GL Hearn (2018) 'Greater Birmingham HMA Strategic Growth Study', https://www.birmingham.gov.uk/downloads/download/1945/greater_birmingham_hma_strategic_growth_study (Accessed: 7/11/2023).

Goode, C. (2023) 'TOD in Regional Urban Growth Boundaries (UGBs): A Case of Transit Adjacent Development or a Strategic Housing Solution?', *Journal of Transport Geography*, 113, pp. 1–10.

Goode, C. (2022A) 'Planning Principles and Particular Places', *Town Planning Review*, 93(3), pp. 301–328.

Goode, C. (2022B) 'The Enduring Importance of Strategic Vision in Planning: The Case of the West Midlands Green Belt', *Planning Perspectives*, 37(6), pp. 1231–1259.

Goodman, R., Freestone, R. and Burton, P. (2017) 'Planning Practice and Academic Research: Views from the Parallel Worlds', *Planning Practice and Research*, 37(4), pp. 1–12.

Gracey, H. (1973) *Containment of Urban England Volume Two* (London: PEP).

Monbiot, G., Grey, R., Kenny, T., Macfarlane, L., Powell-Smith, A., Shrubsole, G. and Stratford, B. (2019) *Land for the Many* (London: Labour Party).

Griffith, M. and Jefferys, P. (2013) *Solutions for the Housing Shortage* (London: Shelter).

Griffiths, S. (2017) *Yes, That's Right: I Hate Green Belt*, (Birmingham: RTPI Tripwire).

Gross, C. (2007) 'Community Perspectives of Wind Energy in Australia: The Application of a Justice and Community Fairness Framework to Increase Social Acceptance', *Energy Policy*, 35(5), pp. 2727–2736.

Grosvenor (2019) 'Rebuilding Trust – Discussion Paper', https://www.grosvenor.com/getattachment/8e97e7a8-e557-4224-bde1-f8833d34acec/Rebuilding-Trust-discussion-paper-(2).pdf (Accessed: 8/08/2022).

Gunn, S. (2007) 'Green Belts: A Review of the Regions' Responses to a Changing Housing Agenda', *Journal of Environmental Planning and Management*, 50(5), pp. 595–616.

Hall, P. (1974) 'The Containment of Urban England', *The Geographical Journal*, 140(3), pp. 386–408.

Hall, P. (2002) *Urban and Regional Planning*. 4th edn. (London: Routledge).

Hall, P. (2014) *Cities of Tomorrow: An Intellectual History of Urban Planning and Design Since 1880* (Oxford: Blackwell Publisher).

Hall, P. (1973) *Containment of Urban England Volume Two* (London: PEP).

Hall, P. and Gracey, H. (1973) *Containment of Urban England Volume Two* (London: PEP).

Hall, P., Gracey, H., Brewett, R. and Thomas, R. (1973A) *Containment of Urban England Volume Two* (London: PEP).

Hall, P., Gracey, H., Brewett, R. and Thomas, R. (1973B) *The Containment of Urban England Volume One* (London: PEP).

Halligan, L. (2019) *Home Truths* (London: Biteback Publishing).

Hamiduddin, I. and Gallent, N. (2024) 'The Rural Housing Market after the Covid-19 Pandemic', *Town Planning Review*, 95(4), pp. 343–354.

Hamnett, C. (2009) 'Spatially Displaced Demand and the Changing Geography of House Prices in London, 1995–2006', *Housing Studies*, 24(3), pp. 301–320.

Harris, J. (2020) 'Plan The World We Need', RTPI, https://www.rtpi.org.uk/new/our-campaigns/plan-the-world-we-need/ (Accessed: 4/1/2024).

Harris, J. (2019) 'The Housing Crisis Is at the Heart of Our National Nervous Breakdown', *The Guardian*, https://www.theguardian.com/commentisfree/2019/oct/28/housing-crisis-houses-brexit-vote (Accessed: 11/12/2023).

Harrison, C. (1981) 'A Playground for Whom? Informal Recreation in London's Green Belt', *Area*, 13(2), pp. 109–114.

Harrison, G. and Clifford, B. (2016) '"The Field of Grain Is Gone; It's Now a Tesco Superstore": Representations of "Urban" and "Rural" Within Historical and Contemporary Discourses Opposing Urban Expansion in England', *Planning Perspectives*, 31(4), pp. 585–609.

Harrison, J., Galland, D. and Tewdwr-Jones, M. (2021) 'Regional Planning Is Dead: Long Live Planning Regional Futures', *Regional Studies*, 55(1), pp. 6–18.

Harvey, D. (1982) *Limits to Capital* (Oxford: Basil Blackwell).

Harvey, D. (1989) 'From Managerialism to Entrepreneurialism: The Transformation of Urban Governance in Late Capitalism', *Geografiska Annaler (Series B)*, 71(1), pp. 3–17.

Harvey, D. (2009) *Social Justice and the City* (Atlanta, GA: University of Georgia Press).

Harvey, D. (2013) *Rebel Cities: From the Right to the City to the Urban Revolution* (New York: Verso Books).

Harvey, W. (2011) 'Strategies for Conducting Elite Interviews', *Qualitative Research*, 11(4), pp. 431–441.

Hatton Parish Plan Steering Group (2013) *Hatton Parish Plan* (Warwick: Hatton).

Haugaard, M. (2021) 'The Four Dimensions of Power: Conflict and Democracy', *Journal of Political Power*, 14(1), pp. 153–175.

Haughton, G. (2017) 'Learning From the Draft Greater Manchester Spatial Framework Process', *Manchester Policy Blog*, http://blog.policy.manchester.ac.uk/posts/2017/09/learning-from-the-draft-greater-manchester-spatial-framework-process/ (Accessed: 3/07/2022).

Haywood, R. (2005) 'Co-Ordinating Urban Development, Stations and Railway Services as a Component of Urban Sustainability: An Achievable Planning Goal in Britain?', *Planning Theory and Practice*, 6(1), pp. 71–97.

Healey, P. (1997) *Collaborative Planning* (Basingstoke: Macmillan).

Healey, P. (2003) 'Collaborative Planning in Perspective', *Planning Theory*, 2(2), pp. 101–123.

Healey, P., Doak, A., McNamara, P. and Elson, M. (1988) *Land Use Planning and the Mediation of Urban Change* (Cambridge: Cambridge University Press).

Heath, S., Charles, V., Crow, G. and Wiles, R. (2007) 'Informed Consent, Gatekeepers and Go-Betweens: Negotiating Consent in Child- and Youth-Orientated Institutions', *British Educational Research Journal*, 33(3), pp. 403–417.

Henley Standard (2020) 'Council Could Lose Control of Plan, Says Minister', https://www.henleystandard.co.uk/news/council/148413/council-could-lose-control-of-local-plan-says-minister.html (Accessed: 4/1/2024).

Herbert Commission (1960) *Royal Commission on Local Government in Greater London* (London: Royal Commission).

Herington, J. (1990) *Beyond Green Belts: Managing Urban Growth in the 21st Century* (London: Regional Studies Association).

Heurkens, E., Adams, D. and Hobma, F. (2015) 'Planners as Market Actors: the Role of Local Planning Authorities in the UK's Urban Regeneration Practice', *Town Planning Review*, 86(6), pp. 625–650.

Highways England (2018) 'A14 Cambridge to Huntingdon', https://assets.highwaysengland.co.uk/roads/road-projects/a14-cambridge-to-huntingdon-improvement/BED20_0026+A14+end+of+scheme+brochure+FINAL.pdf (Accessed: 4/1/2024).

Hilber, C. and Vermeulen, W. (2014) 'The Impact of Supply Constraints on House Prices in England', *Economic Journal*, 126, pp. 358–405.

Hillier, J. (2003) "Agon'izing Over Consensus: Why Habermasian Ideals Cannot Be "real"', *Planning Theory*, 2(1), pp. 37–59.

Hills, V. (2020) 'Open Letter From Chief Executive of the RTPI', RTPI, https://www.rtpi.org.uk/media/5720/open-letter-plantheworldweneed.pdf (Accessed: 21/08/2021).

HM Government (2018) *A Green Future: Our 25 Year Plan to Improve the Environment* (London: DEFRA).

HMSO (1940) *Royal Commission on the Distribution of the Industrial Population* (London: HMSO).

Holman, N., Fernandez-Arrigoitia, M., Scanlon, K. and Whitehead, C. (2015) *Housing in London: Addressing the Supply Crisis* (London: LSE Knowledge Transfer).

Home Owners Alliance (2015) 'Public Support Action to Address Housing Shortage by Advertising New Homes in UK First', https://hoa.org.uk/2015/03/2015-survey-major-housing-policies/ (Accessed: 4/1/2024).

House of Lords (2016) *Building More Homes* (London: House of Lords).

Howard, E. (1946) *Garden Cities of To-Morrow* (London: Faber and Faber LTD).

Howard, E. (2003) *To-Morrow: A Peaceful Path to Reform* (London: Routledge).

Hubbard, P. (2005) '"Inappropriate and Incongruous": Opposition to Asylum Centres in the English Countryside', *Journal of Rural Studies*, 21(1), pp. 3–17.

Hudson, N. and Green, B. (2017) *Missing Movers A Long-Term Decline in Housing Transactions?* (London: Council of Mortgage Lenders).

Hughes, L. and Buffery, S. (2006) *Nottingham-Derby Green Belt Review* (Derbyshire and Nottinghamshire County Councils).

Hurst, D. and Mickan, S. (2017) 'Describing Knowledge Encounters in Healthcare: A Mixed Studies Systematic Review and Development of a Classification', *Implementation Science*, 12(35), pp. 1–14.

Inch, A. (2015) 'Ordinary Citizens and the Political Cultures of Planning: In Search of the Subject of a New Democratic Ethos', *Planning Theory*, 14(4), pp. 404–424.

Inch, A. (2018) '"Opening for Business"? Neoliberalism and the Cultural Politics of Modernising Planning in Scotland', *Urban Studies*, 55(5), pp. 1076–1092.

Inch, A., Dunning, R., While, A., Hickman, H. and Payne, S. (2020) '"The Object Is to Change the Heart and Soul": Financial Incentives, Planning and Opposition to New Housebuilding in England', *Environment and Planning C: Politics and Space*, 38(4), pp. 713–732.

Inch, A., Sartorio, F., Bishop, J., Beebeejaun, Y., McClymont, K., Frediani, A., Cociña, C. and Quick, K. (2019) 'People and Planning at Fifty', *Planning Theory and Practice*, 20(5), pp. 735–759.

Inch, A. and Shepherd, E. (2019) 'Thinking Conjuncturally About Ideology, Housing and English Planning', *Planning Theory*, 19(1), pp. 1–21.

Inch, A., Tait, M. and Chapman, K. (2020) 'A Critical Academic Response to the Evidence-Free Debate on Planning Reform', in *The Wrong Answers to the Wrong Questions* (London: Town and Country Planning Association), pp. 9–13.

Innes, J. and Booher, D. (2015) 'A Turning Point for Planning Theory? Overcoming Dividing discourses', *Planning Theory*, 14(2), pp. 195–213.

Ipsos MORI (2005) '50th Anniversary of Green Belts', https://ems.ipsos-mori.com/researchpublications/researcharchive/435/50th-Anniversary-of-Green-Belts.aspx (Accessed: 4/1/2024).

Ipsos MORI (2015) 'Attitudes Towards Use of Green Belt Land', https://www.ipsos.com/sites/default/files/migrations/en-uk/files/Assets/Docs/Polls/cpre-green-belt-tables-aug-2015.pdf (Accessed: 4/1/2024).

Irvine, A., Drew, P. and Sainsbury, R. (2013) '"Am I Not Answering Your Questions Properly?" Clarification, Adequacy and Responsiveness in Semi-Structured Telephone and Face-to-Face Interviews', *Qualitative Research*, 13(1), pp. 87–106.

James, O. and Lodge, M. (2003) 'The Limitations of "Policy Transfer" and "Lesson Drawing" for Public Policy Research', *Political Studies Review*, 1, pp. 179–193.

Jefferys, P., Lloyd, T., Argyle, A., Sarling, J., Crosby, J. and Bibby, J. (2015) *Building the Homes We Need: A Programme for the 2015 Government* (London: KPMG and Shelter).

Jenks, M., Burton, E. and Williams, K. (2000) *Achieving Sustainable Urban Form* (Abingdon: Land Use Policy).

Jessop, B. (2004) *Spatial Fixes, Temporal Fixes, and Spatio-Temporal Fixes* (Lancaster: University of Lancaster).

Jessop, B. (2018) 'Neoliberalization, Uneven Development, and Brexit: Further Reflections on the Organic Crisis of the British State and Society', *European Planning Studies*, 26(9), pp. 1728–1746.

Kahn, E. (2020) 'Sevenoaks Council 'bemused' at High Court Ruling Over Duty to Cooperate Failure', *Planning Resource*, https://www.planningresource.co.uk/article/1700553/sevenoaks-council-bemused-high-court-ruling-duty-cooperate-failure (Accessed: 4/1/2024).

Katz, C. (1994) 'Playing the Field: Questions of Fieldwork in Geography', *Professional Geographer*, 46(1), pp. 67–72.

Keeble, D. (1971) 'Planning and South East England', *The Royal Geographical Society*, 3(2), pp. 69–74.

Kells, G., Kimber, P., Sullivan, M. and Goode, P. (2007) *What Price West Midlands Green Belts?* (London: CPRE).

Kenny, T. (2019A) 'Land for the Many and a New Politics of Land', *Planning Theory and Practice*, 20(5), pp. 763–768.

Kenny, T. (2019B) 'The UK Planning Profession in 2019', RTPI, https://www.rtpi.org.uk/media/3370677/The Planning Profession in 2019.pdf (Accessed: 4/1/2024).

Kenny, T., Elliott, T. and Bicquelet-Lock, A. (2018) 'Better Planning for Housing Affordability: Three Approaches to Solving the Housing Crisis in the UK', *Journal of Urban Regeneration and Renewal*, 11(3), pp. 233–241.

Kiernan, M. (1983) 'Ideology, Politics, and Planning: Reflections on the Theory and Practice of Urban Planning', *Environment and Planning B*, 10(1), pp. 71–87.

Kilroy, J. (2017) *Better Planning for Housing Affordability* (London: RTPI).

Kirby, M. and Scott, A. (2023) 'Multifunctional Green Belts: A Planning Policy Assessment of Green Belts Wider Functions in England', *Land Use Policy*, 123, pp. 1–14.

Kirby, M., Scott, A., Lugar, J. and Walsh, C. (2023B) 'Beyond Growth Management: A Review of the Wider Functions and Effects of Urban Growth Management Policies', *Landscape and Urban Planning*, 230, pp. 1–12.

Kirby, M., Scott, A. and Walsh, C. (2023A) 'Translating Policy to Place: Exploring Cultural Ecosystem Services in Areas of Green Belt Through Participatory Mapping', *Ecosystems and People*, 19(1), pp. 1–19.

Kirby, M., Scott, A. and Zawadzka, J. (2024) 'Ecosystem Service Multifunctionality and Trade-Offs in English Green Belt Peri-Urban Planning', *Ecosystem Services*, 67, pp. 1–13.

Kjærås, K. (2024) 'The Politics of Urban Densification in Oslo', *Urban Studies*, 61(1), pp. 40–57.

Kraftl, P., Hadfield-Hill, S. and Laxton, A. (2018) *Garden Villages and Towns: Planning for Children and Young People* (Birmingham: University of Birmingham).

Kvale, S. (2007) *Doing Interviews* (London: Sage).

Lainton, A. (2014) 'London's Green Belt Has Never Had a Proper Plan', *Decisions, Decisions, Decisions*, https://andrewlainton.wordpress.com/2014/12/18/londons-green-belt-has-never-had-a-proper-plan/ (Accessed: 4/1/2024).

Lake, R. (1993) 'Planners' Alchemy Transforming NIMBY to YIMBY: Rethinking NIMBY', *Journal of the American Planning Association*, 59(1), pp. 87–93.

Lake, R. (2016) 'Justice As Subject and Object of Planning', *International Journal of Urban and Regional Research*, 40(6), pp. 1205–1220.

Landmark Chambers (2020) 'Have We Got Planning News For You? Episode 10', https://www.youtube.com/watch?v=FeEEt1KJoFA (Accessed: 4/1/2024).

Lane, M. (2019) 'Is NIMBYism on the Rise Again?', *Property Investor Today*, https://www.propertyinvestortoday.co.uk/breaking-news/2019/10/is-nimbyism-on-the-rise-again (Accessed: 4/1/2024).

Law, J. (2004) 'Reviewed Work(s): John Allen', *American Journal of Sociology*, 110(3), pp. 794–79.

Law, M. *et al.* (2000) 'Green Belts in the West Midlands', in Chapman D., Harridge C., Harrison J., Harrison G. and Stokes B. (eds.) *Region and Renaissance: Reflections on Planning and Development in the West Midlands, 1950–2000* (Wrexham: Berwin Books), pp. 50–70.

Layzer, J. (2012) *The Environmental Case*. 3rd edn. (Washington, DC: CQ Press).

Lennon, M. (2017) 'On "the Subject" of Planning's Public Interest', *Planning Theory*, 16(2), pp. 150–168.

Lennon, M. (2020) 'Planning as Justification', *Planning Theory and Practice*, 21(5), pp. 803–807.

Lennon, M. and Fox-Rogers, L. (2017) 'Morality, Power and the Planning Subject', *Planning Theory*, 16(4), pp. 364–383.

Letwin, O. (2018) *Independent Review of Build Out: Draft Analysis* (London: Ministry of Housing, Communities and Local Government).

Lichfields (2016) *Start to Finish: How Quickly Do Large-Scale Housing Sites Deliver?* (London: Lichfields).

Lichfields (2017) *Stock and Flow: Planning Permissions and Housing Output* (London: Lichfields).

Lichfields (2018) *Realising Potential: The Scale and Role of Specialist Land Promoters in Housing Delivery* (London: Lichfields).

Lindsay, J. (2005) 'Getting the Numbers: The Unacknowledged Work in Recruiting for Survey Research', *Field Methods*, 17(1), pp. 119–128.

Litman, T. (2015) 'Analysis of Public Policies That Unintentionally Encourage and Subsidize Urban Sprawl', *The New Climate Economy*, https://newclimateeconomy.report/workingpapers/wp-content/uploads/sites/5/2016/04/public-policies-encourage-sprawl-nce-report.pdf (Accessed: 4/1/2024).

Lloyd, G. (2006) *Planning and the Public Interest in the Modern World* (London: RTPI).

Lloyd, M. and Peel, D. (2007) 'Green Belts in Scotland: Towards the Modernisation of a Traditional Concept?', *Journal of Environmental Planning and Management*, 50(5), pp. 639–656.

London First/Quod (2015) *The Green Belt: A Place for Londoners* (London: Quod).

Longhurst, R. *et al.* (2010) 'Semi-Structured Interviews and Focus groups', in Clifford N., French S. and Valentine G. (eds.) *Key Methods in Human Geography* (London: Sage), pp. 117–132.

Longley, P., Batty, M., Shepherd, J. and Sadler, G. (1992) 'Do Green Belts Change the Shape of Urban Areas? A Preliminary Analysis of the Settlement Geography of South East England', *Regional Studies*, 26(5), pp. 437–452.

Lord, A. and Tewdwr-Jones, M. (2018) 'Getting the Planners Off Our Backs: Questioning the Post-Political Nature of English Planning Policy', *Planning Practice and Research*, 34(3), pp. 1–15.

Low, S. (2013) 'Public Space and Diversity: Distributive, Procedural and Interactional Justice for Parks', in Young G. and Stevenson D. (eds.) *The Ashgate Research Companion to Planning and Culture* (Surrey: Ashgate Publishing), pp. 295–310.

Lowe, K. (2020) *Key High Court Ruling* (Cirencester: Pegasus Group).

Lowndes, V. and Pratchett, L. (2012) 'Local Governance Under the Coalition Government: Austerity, Localism and the "Big Society"', *Local Government Studies*, 38(1), pp. 21–40.

Lukes, S. (2004) *Power: A Radical View*. 2nd edn. (Basingstoke: Palgrave).

Lund, B. (2017) *Housing Politics in the United Kingdom* (Bristol: Policy Press).

Lund, B. (2019) *Housing in the United Kingdom: Whose Crisis?* (Basingstoke: Palgrave Macmillan).

Lyons, M. (2014) *The Lyons Housing Review* (London: The Labour Party).

Mace, A. (2017) 'Beware of the New Justifications for the Green Belt: What We Need Is a New Approach', *LSE British Politics and Policy*, http://blogs.lse.ac.uk/politicsandpolicy/beware-new-justifications-for-green-belt-what-we-need-is-a-new-approach/ (Accessed: 4/1/2024).

Mace, A. (2018) 'The Metropolitan Green Belt, Changing An Institution', *Progress in Planning*, 121, pp. 1–28.

Mace, A., Blanc, F., Gordon, I. and Scanlon, K. (2016) *A 21st Century Metropolitan Green Belt* (London: LSE Knowledge Exchange).

Maidment, C. (2016) 'In the Public Interest? Planning in the Peak District National Park', *Planning Theory*, 15(4), pp. 366–385.

Malik, I. (2024) 'Can Political Ecology Be Decolonised? A Dialogue With Paul Robbins', *Transactions of the Institute of British Geographers*, 11(1), pp. 1–13.

Manns, J. (2014) *Green Sprawl: Our Current Affection for a Preservation Myth?* (London: London Society).

Manns, J. and Falk, N. (2016) *Re/Shaping London* (London: London Society).

Marais, L., Denoon-Stevens, S. and Cloete, J. (2020) 'Mining Towns and Urban Sprawl in South Africa', *Land Use Policy*, 93, pp. 1–12.

Marsh, D. and McConnell, A. (2010) 'Towards a Framework for Establishing Policy Success', *Public Administration*, 88(2), pp. 564–583.

Marston, S., Jones, J. and Woodward, K. (2005) 'Human Geography Without Scale', *Transactions of the Institute of British Geographers*, 30(4), pp. 416–432.

Martin, D. (2024) 'Starmer Pledges to 'take the brakes off' Planning Rules to Build 1.5m Homes', *The Telegraph*, https://www.telegraph.co.uk/politics/2024/07/17/labour-to-build-15m-homes-starmer-kings-speech/, (Accessed: 25/07/2024).

Martin, R. (2001) 'Geography and Public Policy: The Case of the Missing Agenda', *Progress in Human Geography*, 25(2), pp. 189–210.

Martin, R. (2010) 'Roepke Lecture in Economic Geography – Rethinking Regional Path Dependence: Beyond Lock-in to Evolution', *Economic Geography*, 86(1), pp. 1–27.

Mason, J. (2018) *Qualitative Researching* (London: Sage).

Mason, J. (2006A) 'Mixing Methods in a Qualitatively Driven Way', *Qualitative Research*, 6(1), pp. 9–25.

Mason, J. (2006B) 'Six Strategies for Mixing Methods and Linking Data in Social Science Research', *NCRM Working Paper Series*, 4(06), pp. 1–14. http://eprints.ncrm.ac.uk/482/ (Accessed: 02/04/2024).

Mason, R. (2023) 'Sunak Stands by Pledge to Protect Green Belt After Starmer Housing Comments', *The Guardian*, https://www.theguardian.com/politics/2023/may/18/rishi-sunak-stands-by-pledge-to-protect-green-belt-after-starmer-housing-comments (Accessed: 02/04/2024).

Massey, D. (1994) *Space, Place and Gender* (Cambridge: Policy Press).

Matless, D. (1998) *Landscape and Englishness* (London: Reaktion Books).

Matthews, P., Bramley, G. and Hastings, A. (2015) 'Homo Economicus in a Big Society: Understanding Middle-Class Activism and NIMBYism Towards New Housing Developments', *Housing, Theory and Society*, 32(1), pp. 54–72.

May, T. (2011) *Social Research: Issues, Methods and Processes* (Milton Keynes: Open University Press).

McAllister, P. (2017) 'The Calculative Turn in Land Value Capture: Lessons from the English Planning System', *Land Use Policy*, 63, pp. 122–129.

McAllister, P., Street, E. and Wyatt, P. (2016) 'Governing Calculative Practices: An Investigation of Development Viability Modelling in the English Planning System', *Urban Studies*, 53(11), pp. 2363–2379.

McClymont, K. (2011) 'Revitalising the Political: Development Control and Agonism in Planning Practice', *Planning Theory*, 10(3), pp. 239–256.

McConnell, A. (2010) *Understanding Policy Success: Rethinking Public Policy* (Basingstoke: Palgrave Macmillan).

McCrum, D. (2014) 'Let's Nationalise the Green Belt', *Financial Times Alphaville*, https://ftalphaville.ft.com/2014/06/06/1750042/lets-nationalise-the-green-belt/ (Accessed: 4/1/2024).

McDowell, L. (2016) 'Reflections on Feminist Economic Geography: Talking to Ourselves?', *Environment and Planning A*, 48(10), pp. 2093–2099.

McGuinness, D., Greenhalgh, P. and Grainger, P. (2018) 'Does One Size Fit All?', *Local Economy*, 33(3), pp. 329–346.

McGuinness, D. and Mawson, J. (2017) 'The Rescaling of Sub-National Planning: Can Localism Resolve England's Spatial Planning Conundrum?', *Town Planning Review*, 88(3), pp. 283–303.

McKee, K., Muir, J. and Moore, T. (2017) 'Housing Policy in the UK: The Importance of Spatial Nuance', *Housing Studies*, 32(1), pp. 60–72.

McLoughlin, J. (1985) *The Systems Approach to Planning: A Critique* (Hong Kong: University of Hong Kong).

Meen, G. (2018) 'Policy Approaches for Improving Affordability', *UK Collaborative Centre for Housing Evidence*, https://thinkhouse.org.uk/site/assets/files/1566/policy.pdf (Accessed: 4/1/2024).

Meen, G. and Whitehead, C. (2020) *Understanding Affordability the Economics of Housing Markets* (Bristol: Policy Press).

Mell, I. (2024) 'Viewpoint – Examining the Long-Term Impacts of Covid-19 on Green Infrastructure: Reflections on Changes in Public Perceptions and Government Action', *Town Planning Review*, 95(4), pp. 363–378.

Menotti, V. (2005) 'Review: The New Transit Town: Best Practices in Transit-Orientated Development', *American Planning Association*, 71(1), p. 111.

Mensah, C., Andres, L., Perera, U. and Roji, A. (2016) 'Enhancing Quality of Life Through the Lens of Green Spaces: A Systematic Review Approach', *International Journal of Wellbeing*, 6(1), pp. 142–163.

MHCLG (2018) *Public Attitudes to Housebuilding: Findings from the 2017 British Social Attitudes Survey* (London: Ministry of Housing, Communities and Local Government).

MHCLG (2019) 'Local Planning Authority Green Belt: England 2018/19', *Planning Statistical Release*, https://assets.publishing.service.gov.uk/government/uploads/system/uploads/attachment_data/file/856100/Green_Belt_Statistics_England_2018-19.pdf (Accessed: 4/1/2024).

MHCLG (2019) 'Public Attitudes to Social Housing: Findings from the 2018 British Social Attitudes Survey', https://assets.publishing.service.gov.uk/government/uploads/system/uploads/attachment_data/file/818535/Public_attitudes_towards_social_housing.pdf (Accessed: 4/1/2024).

MHCLG (2020A) *Planning for the Future* (London: Ministry of Housing, Communities and Local Government).

MHCLG (2020B) 'Transparency and Competition: A Call for Evidence on Data on Land Control, Ministry of Housing Communities and Local Government', https://assets.publishing.service.gov.uk/government/uploads/system/uploads/attachment_data/file/907213/Call_for_evidence_on_Contractual_Controls.pdf (Accessed: 4/1/2024).

MHCLG (2024) *English Devolution White Paper* (London: MHCLG).

Miller, M. (1992) *Raymond Unwin: Garden Cities and Town Planning* (Leicester: Leicester University Press).

Ministry of Housing, Communities and Local Government (2024) National Planning Policy Framework, https://assets.publishing.service.gov.uk/media/675abd214cbda57cacd3476e/NPPF-December-2024.pdf (Accessed 18/12/2024).

Ministry of Housing and Local Government (1955) *Circular 42/55 – Green Belts* (London: HMSO).

Ministry of Housing and Local Government (1957) *Green Belts, Circular No. 50/57* (London: HMSO).

Ministry of Works and Planning (1942) *The Scott Report* (London: HMSO).

Mitchell, L. *et al.* (1999) 'Combining Focus Groups and Interviews: Telling How It Is; Telling How It feels', in Barbour R.S. and Kitzinger J. (eds.) *Developing Focus Group Research: Politics, Theory and Practice* (London: Sage), pp. 36–46.

Modell, S. (2009) 'In Defence of Triangulation: A Critical Realist Approach to Mixed Methods Research in Management Accounting', *Management Accounting Research*, 20(3), pp. 208–221.

Monk, S. and Whitehead, C. (1999) 'Evaluating the Economic Impact of Planning Controls in the United Kingdom: Some Implications for Housing', *Land Economics*, 75(1), pp. 74–93.

Montell, F. (2017) 'Focus Group Interviews : A New Feminist Method', *NWSA Journal*, 11(1), pp. 44–71.

Morphet, J. (2011) *Effective Practice in Spatial Planning* (Abingdon: Routledge).

Morphet, J. and Clifford, B. (2019) 'Local Authority Direct Delivery of Housing: Continuation Research', *RTPI Research Paper*, https://www.rtpi.org.uk/media/2043/local-authority-direct-delivery-of-housing-ii-continuation-research-full-report.pdf (Accessed: 4/1/2024).

Morris, H. (2019) 'NIMBYism Spurs 1.9 Million Objections Since 2017', *The Planner*, https://www.theplanner.co.uk/news/nimbyism-spurs-19-million-objections-since-2017 (Accessed: 4/1/2024).

Morrison, N. (2010) 'A Green Belt Under Pressure: The Case of Cambridge, England', *Planning Practice and Research*, 25(2), pp. 157–181.

Morriss, P. (2002) *Power: A Philosophical Analysis*. 2nd edn. (New York: Manchester University Press).

Mouffe, C. (2000) 'Deliberative Democracy or Agonistic Pluralism?', *Social Research*, 66(3), pp. 745–758.

Mulheirn, I. (2019) 'Tackling the UK Housing Crisis: Is Supply the Answer?', *UK Collaborative Center for Housing Evidence*, https://housingevidence.ac.uk/wp-content/uploads/2024/05/20190820b-CaCHE-Housing-Supply-FINAL.pdf (Accessed: 4/1/2024).

Munro, M. (2018) 'House Price Inflation in the News: A Critical Discourse Analysis of Newspaper Coverage in the UK', *Housing Studies*, 33(7), pp. 1085–1105.

Munton, R. (1983) *London's Green Belt: Containment in Practice* (London: Allen and Unwin).

Munton, R. (1986) 'Green Belts: The End of an Era?', *Geography*, 71(3), pp. 206–214.

Munton, R., Whatmore, S. and Marsden, T. (1988) 'Reconsidering Urban-Fringe Agriculture: A Longitudinal Analysis of Capital Restructuring on Farms in the Metropolitan Green Belt', *Transactions of the Institute of British Geographers*, 13(3), pp. 324–336.

Myers, J. (2017) *Yes In My Back Yard: How To End The Housing Crisis, Boost The Economy And Win More Votes* (London: Adam Smith Institute).

Nadin, V. and Stead, D. (2008) 'European Spatial Planning Systems, Social Models and Learning', *disP The Planning Review*, 44(172), pp. 35–47.

Nagar, R., Lawson, V., McDowell, L. and Hanson, S. (2009) 'Locating Globalisation: Feminist (Re)Readings of the Subjects and Spaces of Globalisation', *Economic Geography*, 78(3), pp. 257–284.

Natarajan, L. (2019) 'Perspectives on Scale in Participatory Spatial Planning', *Built Environment*, 45(2), pp. 230–247.

NatCen Social Research (2017) British Social Attitudes Survey, 2016, UK Data Service. doi: SN: 8252.

Nathan, M. (2007) 'A Question of Balance: Cities, Planning and the Barker Review', *Centre for Cities Discussion Paper 9*, pp. 1–12.

Natural England (2007) *Natural England's Policy Position on Housing Growth and Green Infrastructure: Pre-Scoping Paper on Principles* (Sheffield: Natural England).

Neuman, M. (2005) 'The Compact City Fallacy', *Journal of Planning Education and Research*, 25(1), pp. 11–26.

Neuman, W. (2011) *Social Research Methods – Qualitative and Quantitative Approaches* (Seventh. Harlow: Pearson).

O'Callaghan, C. and McGuirk, P. (2021) 'Situating Financialisation in the Geographies of Neoliberal Housing Restructuring: Reflections from Ireland and Australia', *Environment and Planning A*, 53(4), pp. 809–827.

O'Reilly, M. and Parker, N. (2013) '"Unsatisfactory Saturation": A Critical Exploration of the Notion of Saturated Sample Sizes in Qualitative Research', *Qualitative Research*, 13(2), pp. 190–197.

O'Toole, R. (2007A) 'Debunking Portland: The City That Doesn't Work', *Policy Analysis*, 596, pp. 1–24.

O'Toole, R. (2007B) *The Best-Laid Plans: How Government Planning Harms Your Quality of Life, Your Pocketbook and Your Future* (Washington DC: Cato Institute).

Olesen, K. (2018) 'Teaching Planning Theory as Planner Roles in Urban Planning Education', *Higher Education Pedagogies*, 3(1), pp. 302–318.

Oliveira, F. (2014) 'Green Wedges: Origins and Development in Britain', *Planning Perspectives*, 29(3), pp. 357–379.

Oliveira, F. (2017) *Green Wedge Urbanism: History, Theory and Contemporary Practice* (London: Bloomsbury Academic).

Onwuegbuzie, A. and Leech, N. (2005) 'Taking the "q" Out of Research: Teaching Research Methodology Courses Without the Divide between Quantitative and Qualitative Paradigms', *Quality and Quantity*, 39(3), pp. 267–296.

Ormerod, E. and MacLeod, G. (2019) 'Beyond Consensus and Conflict in Housing Governance: Returning to the Local State', *Planning Theory*, 18(3), pp. 319–338.

Palfrey, C., Thomas, P. and Philips, C. (2012) *Evaluation for the Real World* (Bristol: Policy Press).

Papa, E. and Bertolini, L. (2015) 'Accessibility and Transit-Orientated Development in European Metropolitan Areas', *Journal of Transport Geography*, 47, pp. 7–83.

Papworth, T. (2015) *The Green Noose: An Analysis of Green Belts and Proposals for Reform* (London: Adam Smith Institute).

Parham, S. and Boyfield, K.R. (2016) *Garden Cities – Why Not?* (London: International Garden Cities Institute).

Park, A., Clery, E., Curtice, J., Phillips, M. and Utting, D. (2012) *British Social Attitudes* (London: NatCen Social Research).

Parker, G., Lynn, T. and Wargent, M. (2015) 'Sticking to the Script? The Co-Production of Neighbourhood Planning in England', *Town Planning Review*, 86(5), pp. 519–536.

Parker, G. and Street, E. (2015) 'Planning at the Neighbourhood Scale: Localism, Dialogic Politics, and the Modulation of Community Action', *Environment and Planning C: Government and Policy*, 33(4), pp. 794–810.

Parker, G., Street, E. and Wargent, M. (2018) 'The Rise of the Private Sector in Fragmentary Planning in England', *Planning Theory and Practice*, 19(5), pp. 734–750.

Parker, G., Street, E. and Wargent, M. (2020) 'Public-Private Entanglements: Consultant Use by Local Planning Authorities in England', *European Planning Studies*, 28(1), pp. 192–210.

Parkes, T. (2019) 'Mayor Andy Street Criticises 'easy option' of Building on Black Country Green Belt', *Express and Star*, https://www.expressandstar.com/news/local-hubs/2019/12/31/andy-street-opposes-report-highlighting-homes-needed-on-green-belt/ (Accessed: 31 December 2019).

Payne, S. (2020) 'Advancing Understandings of Housing Supply Constraints: Housing Market Recovery and Institutional Transitions in British Speculative Housebuilding', *Housing Studies*, 35(2), pp. 266–289.

Peck, J. (2015) 'Cities Beyond Compare?', *Regional Studies*, 49(1), pp. 160–182.

Peck, J. and Tickell, A. (1996) 'The Return of the Manchester Men', *Institute of British Geographers, New Series*, 21(4), pp. 595–616.

Peck, J. and Tickell, A. (2003) 'Making Global Rules: Globalisation or Neoliberalism?', in Peck J. and Yeung H. (eds) *Remaking the Global Economy: Economic Geographical Perspectives* (Basingstoke: SAGE Publications), pp. 163–182.

Peck, J. and Tickell, A. (2012) 'Apparitions of Neoliberalism: Revisiting "Jungle Law Breaks Out"', *Area*, 44(2), pp. 245–249.

Pendlebury, D. (2015) *What Critics of Planning Should Understand About Solving the Housing Crisis* (London: RTPI).

Pendlebury, J. (2001) 'Alas Smith and Burns? Conservation in Newcastle upon Tyne City Centre 1959–1968', *Planning Perspectives*, 16(2), pp. 115–141.

Pennington, M. (2000) *Planning and the Political Market: Public Choice and the Politics of Government Failure* (London: Athlone Press).

Peter Brett Associates (2015) 'Strategic Housing Needs Study: Stage 3 Report', http://centreofenterprise.com/wp-content/uploads/2015/09/SHNS-Phase-3.pdf (Accessed: 4/1/2024).

Pike, A., Coombes, M., O'Brien, P. and Tomaney, J. (2018) 'Austerity States, Institutional Dismantling and the Governance of Sub-National Economic Development: The Demise of the Regional Development Agencies in England', *Territory, Politics, Governance*, 6(1), pp. 118–144.

Plaw, A. (2007) 'Luke's Three-Dimensional Modelling of Power Redux: Is It Still Compelling?', *Social Theory and Practice*, 33(3), pp. 489–500.

Pløger, J. (2004) 'Strife: Urban Planning and Agonism', *Planning Theory*, 3(1), pp. 71–92.

Power, A. and Haughton, J. (2007) *Jigsaw Cities: Big Places, Small Spaces* (Bristol: Policy Press).

Prior, A. and Raemaekers, J. (2007) 'Is Green Belt Fit for Purpose in a Post-Fordist Landscape?', *Planning Practice and Research*, 22(4), pp. 579–599.

Pykett, J. (2022) 'Spatialising Happiness Economics: Global Metrics, Urban Politics, and Embodied Technologies', *Transactions of the Institute of British Geographers*, 47, pp. 635–650.

Quinn, C. (2024) 'Homes on Green Belt: Rachel Reeves Sets Out Homebuilding Plans to Boost Growth in First Speech as Chancellor', *LBC*, https://www.lbc.co.uk/news/greenbelt-homes-rachel-reeves-chancellor/ (Accessed: 08/07/2024).

Raco, M., Durrant, D. and Livingstone, N. (2018) 'Slow Cities, Urban Politics and the Temporalities of Planning: Lessons from London', *Environment and Planning C: Politics and Space*, 36(7), pp. 1176–1194.

Raco, M., Durrant, D. and Livingstone, N. (2019) 'Seeing Like an Investor: Urban Development Planning, Financialisation, and Investors' Perceptions of London as an Investment Space', *European Planning Studies*, 27(6), pp. 1064–1082.

Rae, A. (2016) 'The 8 English Regions of a Federal UK', *Stats, Maps and Pix*, http://www.statsmapsnpix.com/2016/12/the-8-english-regions-of-federal-uk.html (Accessed: 7/8/2019).

Rankl, F., Barton, C. and Carthew, H. (2023) *Green Belt* (London: House of Commons Library).

Raynsford Review (2018A) *Planning 2020: Interim Report of the Raynsford Review* (London: Town and Country Planning Association).

Raynsford Review (2018B) *Planning 2020: Raynsford Review of Planning in England* (London: Town and Country Planning Association).

Rees-Mogg, J. and Tylecote, R. (2019) 'Raising The Roof', *Institute of Economic Affairs*, file:///C:/Users/cxg227/Downloads/CC70_Raising-the-roof_web (1).pdf (Accessed: 4/1/2024).

Reyes-Riveros, R., Altamirano, A., De La Barrera, F., Rozas-Vásquez, D., Vieli, L. and Meli, P. (2021) 'Linking Public Urban Green Spaces and Human Well-Being: A Systematic Review', *Urban Forestry and Urban Greening*, 61, pp. 1–15.

Riddell, C. (2020) 'Planning Reforms and the Role of Strategic Planning', *Catriona Riddell Associates*, file:///C:/Users/Charles/Downloads/Catriona-Riddell-Planning-Reforms-the-Role-of-Strategic-Planning-2.pdf (Accessed: 4/1/2024).

Riley-Smith, B., Fisher, L. and Hope, C. (2021) 'Boris Johnson Pledges No Homes to be Built on Green Fields', *The Telegraph*, https://www.telegraph.co.uk/politics/2021/10/06/boris-johnson-pledges-no-homes-green-fields/ (Accessed: 03/04/2024).

Roberts, H. (2017) 'Using Twitter Data in Urban Green Space Research: A Case Study and Critical Evaluation', *Applied Geography*, 81, pp. 13–20.

Robinson, N. (2006) 'Learning from Lukes?: The Three Faces of Power and the European Union', *ECPR*, https://ecpr.eu/filestore/paperproposal/b40088b9-6ac4-44cf-90da-ee843ce5f1a0.pdf, (Accessed: 4/1/2024).

Roy, S. (2011) 'Politics, Passion and Professionalization in Contemporary Indian Feminism', *Sociology*, 45(4), pp. 587–602.

RTPI (2015) *Building in the Green Belt? A Report into Commuting Patterns in the Metropolitan Green Belt* (London: RTPI).

RTPI (2016A) *Where Should We Build New Homes? RTPI Policy Statement* (London: RTPI).

RTPI (2016B) *The Location of Development: Mapping Planning Permissions in Housing in Twelve English City-Regions* (London: RTPI).

RTPI (2024) *The Location of Development 4* (London: RTPI).

Russell, J., Fudge, N. and Greenhalgh, T. (2020) 'The Impact of Public Involvement in Health Research', *Research Involvement and Engagement*, 6(63), pp. 1–8.

Ryan-Collins, J. (2018) *Why Can't You Afford a Home?* (Bristol: Polity Press).

Ryan-Collins, J. (2021) 'Breaking the Housing–Finance Cycle: Macroeconomic Policy Reforms for More Affordable Homes', *Environment and Planning A*, 53(3), pp. 480–502.

Rydin, Y. (1985) 'Residential Development and the Planning System: A Study of the Housing Land System at the Local Level', *Progress in Planning*, 24(1), pp. 3–69.

Rydin, Y. (2011) *The Purpose of Planning: Creating Sustainable Towns and Cities* (Bristol: Policy Press).

Rydin, Y. (2013) *The Future of Planning: Beyond Growth Dependence* (Bristol: Policy Press).

Rydin, Y. (2020) 'Silences, Categories and Black Boxes', *Planning Theory*, 19(2), pp. 214–233.

Rydin, Y. and Pennington, M. (2000) 'Public Participation and Local Environmental Planning', *Local Environment: The International Journal of Justice and Sustainability*, 5(2), pp. 153–169.

Sandercock, L. (1998) *Towards Cosmopolis: Planning for Multicultural Cities* (Chichester: John Wiley).

Sandercock, L. and Dovey, K. (2002) 'Pleasure, Politics, and the "Public Interest": Melbourne's Riverscape Revitalization', *Journal of the American Planning Association*, 68(2), pp. 151–164.

Satsangi, M. and Dunmore, K. (2003) 'The Planning System and the Provision of Affordable Housing in Rural Britain: A Comparison of the Scottish and English Experience', *Housing Studies*, 18(2), pp. 201–217.

Saunders, P. (2016) *Restoring a Nation of Home Owners* (London: Civitas).

Saunders, M., Lewis, P. and Thornhill, A. (2016) *Research Methods for Business Students*. 7th edn. (Harlow: Pearson).

Saunders, M. and Townsend, K. (2016) 'Reporting and Justifying the Number of Interview Participants in Organization and Workplace Research', *British Journal of Management*, 27(4), pp. 836–852.

Scapens, R. *et al.* (2004) 'Doing Case Study Research', in Humphrey C. and Lee B. (eds.) *The Real Life Guide to Accounting Research* (London: Elsevier), pp. 257–281.

ScotCen Social Research (2015) 'Scottish Social Attitudes Survey, 2013, UK Data Service. SN: 7519', http://doi.org/10.5255/UKDA-SN-7519-2 (Accessed: 4/1/2024).

Scott, A.J., Carter, C., Reed, M.R., Larkham, P., Adams, D., Morton, N., Waters, R., Collier, D., Crean, C., Curzon, R., Forster, R., Gibbs, P., Grayson, N., Hardman, M., Hearle, A., Jarvis, D., Kennet, M., Leach, K., Middleton, M., Schiessel, N., Stonyer, B. and Coles, R. (2013) 'Disintegrated Development at the Rural-Urban Fringe: Re-Connecting Spatial Planning Theory and Practice', *Progress in Planning*, 83, pp. 1–52.

Scott, M. (2015) 'Rebuilding Britain Planning for a Better Future', *Planning Theory and Practice*, 16(1), pp. 139–141.

Self, P. *et al.* (1962) 'Introduction – The Green Belt Idea', in Long J. (ed.) *The Wythall Inquiry* (London: Town and Country Planning Association), pp. vii–xxi.

Shaw, T. (2007) 'Editorial', *Journal of Environmental Planning and Management*, 50(5), pp. 575–578.

Shelter and Quod (2016) *When Brownfield isn't Enough: Strategic Options for London's Growth* (London: Quod/Shelter).

Short, J., Fleming, S. and Witt, S. (1987) 'Conflict and Compromise in the Built Environment: Housebuilding in Central Berkshire', *Transactions of the Institute of British Geographers*, 12(1), pp. 29–42.

Shrubsole, G. (2019) *Who Owns England* (Glasgow: William Collins).

Silva, E., Healey, P., Harris, N. and Van den Broeck, P. (eds.) (2014) *The Routledge Handbook of Planning Research Methods* (London: Routledge).

Simmie, J. (1981) *Power, Property and Corporatism* (London: Macmillan).

Simons, Z. (2020) 'We Need to Talk About England's Greenbelt', *Financial Times*, https://www.ft.com/content/f240e825-eca0-4fba-8813-86dfaa0a6933 (Accessed: 15/12/2020).

Sims, S. and Bossetti, N. (2016) *Stopped: Why People Oppose Residential Development in Their Back Yard* (London: Centre for London).

Slade, D., Gunn, S. and Schoneboom, A. (2019) *Serving the Public Interest* (London: RTPI).

Smith, N. (1996) *The New Urban Frontier: Gentrification and the Revanchist City* (London: Routledge).

Soja, E. (2010) *Seeking Spatial Justice* (Minneapolis, MN: University of Minnesota Press).

Solihull Metropolitan Borough Council (2020) Solihull Local Plan – Draft Submission Plan, https://www.solihull.gov.uk/sites/default/files/2020-12/Draft-Submission-Plan-Oct-2020%20(1).pdf (1).pdf (Accessed: 4/1/2024).

Sorensen, A. (2002) *The Making of Urban Japan* (London: Routledge).

Sorensen, A. (2015) 'Taking Path Dependence Seriously: An Historical Institutionalist Research Agenda in Planning History', *Planning Perspectives*, 30(1), pp. 17–38.

Squires, G. (2022) *The Economics of Property and Planning: Future Value* (Abingdon: Routledge).

Squires, G. and Heurkens, E. (eds.) (2014) *International Approaches to Real Estate Development* (Abingdon: Routledge).

Stafford, S. (2021) 'YIMBYs and NIMBYs. Is Planning Becoming a New Front in the Culture War?', *50 Shades of Planning*, https://samuelstafford.blogspot.com/2021/06/yimbys-versus-nimbys-is-planning-new.html (Accessed: 5/4/2024).

Stringer, B. (2015) 'Is the Green Belt sustainable?', *Barney's Blog*, https://barneystringer.wordpress.com/2014/06/17/is-the-green-belt-sustainable/ (Accessed: 4/1/2024).

Sturzaker, J. (2010) 'The Exercise of Power to Limit the Development of New Housing in the English Countryside', *Environment and Planning A*, 42(4), pp. 1001–1016.

Sturzaker, J. (2011) 'Can Community Empowerment Reduce Opposition to Housing? Evidence from Rural England', *Planning Practice and Research*, 26(5), pp. 555–570.

Sturzaker, J. (2017) 'How the Planning System Lets Homeowners Overwhelm the Broader Public Interest', *Democratic Audit UK*, http://www.democraticaudit.com/ (Accessed: 5/1/2018).

Sturzaker, J. and Mell, I. (2016) *Green Belts: Past, Present and Future* (Abingdon: Routledge).

Sturzaker, J. and Shucksmith, M. (2011) 'Planning for Housing in Rural England: Discursive Power and Spatial Exclusion', *Town Planning Review*, 82(2), pp. 169–194.

Tait, M. (2016) 'Planning and the Public Interest: Still a Relevant Concept for Planners?', *Planning Theory*, 15(4), pp. 335–343.

Tait, M. and Campbell, H. (2000) 'The Politics of Communication between Planning Officers and Politicians: The Exercise of Power Through Discourse', *Environment and Planning A*, 32(3), pp. 489–506.

Tait, M. and Inch, A. (2016) 'Putting Localism in Place: Conservative Images of the Good Community and the Contradictions of Planning Reform in England', *Planning Practice and Research*, 31(2), pp. 174–194.

Tallon, A. (2013) *Urban Regeneration in the UK* (London: Routledge).

Taylor, M. (2015) *Garden Villages: Empowering Localism to Solve the Housing Crisis* (London: Policy Exchange).

TCPA (2002) *Policy Statement: Green Belts* (London: TCPA).

Telegraph Hill Society (2018) Submission on Matter M65 (Green Belt and Metropolitan Open Land), https://www.london.gov.uk/sites/default/files/m65_telegraph_hill_society_2315.pdf (Accessed: 4/1/2024).

Tewdwr-Jones, M. and Allmendinger, P. (1998) 'Deconstructing Communicative Rationality: A Critique of Habermasian Collaborative Planning', *Environment and Planning A*, 30(11), pp. 1975–1989.

The Economist (2014) 'Yes, Planning Policy is the Cause of the Housing Crisis', *The Economist*, https://www.economist.com/blogs/blighty/2014/08/housebuilding-britain (Accessed: 4/1/2024).

The Economist (2017) 'Philip Hammond's Budget Spots Britain's Problems But Fails to Fix Them', *The Economist*, https://www.economist.com/news/britain/21731619-chancellors-cautious-budget-shows-how-he-hemmed-economically-and-politically-philip?zid=310&ah=4326ea44f22236ea534e2010ccce1932 (Accessed: 4/1/2024).

The Economist (2024) 'Would Building 1.5m Homes Bring Down British House Prices?', https://www.economist.com/britain/2024/08/08/would-building-15m-homes-bring-down-british-house-prices (Accessed: 15/8/2024).

Thomas, D. (1963) 'London's Green Belt: The Evolution of an Idea', *The Geographical Journal*, 129(1), pp. 14–24.

Thrift, N. (2004) 'Intensities of Feeling: Towards a Spatial Politics of Affect', *Geografiska Annaler, Series B: Human Geography*, 86(1), pp. 57–78.

Tickell, A. (1995) 'Reflections on "Activism and the Academy"', *Environment and Planning D: Society and Space*, 13(2), pp. 235–237.

Tracy, S.J. (2010) 'Qualitative Quality: Eight a "big-Tent" Criteria for Excellent Qualitative Research', *Qualitative Inquiry*, 16(10), pp. 837–851.

UK2070 Commission (2020) '"Make No Little Plans': Acting at Scale for a Fairer and Stronger Future', https://uk2070.org.uk/wp-content/uploads/2022/06/UK2070-FINAL-REPORT-Copy.pdf (Accessed: 4/1/2024).

Underdown, D. (1985) *Revel, Riot and Rebellion* (Oxford: Oxford University Press).

Upton, W. (2019) 'What Is the Purpose of Planning Policy? Reflections on the Revised National Planning Policy Framework 2018', *Journal of Environmental Law*, 31(1), pp. 135–149.

Urbed (2014) 'Wolfson Economics Prize 2014', *Urbed*, http://urbed.coop/wolfson-economic-prize (Accessed: 4/1/2024).

Valler, D., Jonas, A. and Robinson, L. (2023) 'Evaluating Regional Spatial Imaginaries: The Oxford–Cambridge Arc', *Territory, Politics, Governance*, 11(3), pp. 434–455.

Valler, D. and Phelps, N.A. (2018) 'Framing the Future: On Local Planning Cultures and Legacies', *Planning Theory and Practice*, 19(5), pp. 1–19.

van den Nouwelant, R., Davison, G., Gurran, N., Pinnegar, S. and Randolph, J. (2015) 'Delivering Affordable Housing Through the Planning System in Urban Renewal Contexts: Converging Government Roles in Queensland, South Australia and New South Wales', *Australian Planner*, 52(2), pp. 77–89.

Vigar, G., Gunn, S. and Brooks, E. (2017) 'Governing Our Neighbours: Participation and Conflict in Neighbourhood Planning', *Town Planning Review*, 88(4), pp. 423–442.

Walker, J. (2016) 'Tory Andrew Mitchell Condemns Communities Secretary Sajid Javid Over Plan for 6,000 Green Belt Homes', *Birmingham Post*, https://www.birminghammail.co.uk/news/midlands-news/andrew-mitchell-condemns-sajid-javid-12290502 (Accessed: 4/1/2024).

Walker, P. (2007) 'Political Ecology: Where Is the Politics?', *Progress in Human Geography*, 31(3), pp. 363–369.

Wang, X., Squires, G. and Dyason, D. (2024) 'Parental Financial Support for Housing: The Importance of Investment Homes and Family Size', *International Journal of Housing Markets and Analysis*, 11, pp. 1–28.

Wannop, U. and Cherry, G. (1994) 'The Development of Regional Planning in the United Kingdom', *Planning Perspectives*, 9(1), pp. 29–60.

Ward, S. (2004) *Planning and Urban Change* (Sage: London).

Ward, K. (2005A) '"Geography and Public Policy: A Recent History of 'policy relevance'", *Progress in Human Geography*, 29(3), pp. 310–321.

Ward, S. (2005B) 'Consortium Developments Ltd and the Failure of "new Country Towns" in Mrs Thatcher's Britain', *Planning Perspectives*, 20(3), pp. 329–359.

Wargent, M. and Parker, G. (2018) 'Re-Imagining Neighbourhood Governance: The Future of Neighbourhood Planning in England', *Town Planning Review*, 89(4), pp. 389–402.

Watling, S. and Breach, A. (2023) 'The Housebuilding Crisis: The UK's 4 Million Missing Homes', *Centre for Cities*, https://www.centreforcities.org/publication/the-housebuilding-crisis/ (Accessed: 01/12/2024).

Warren, C. and Clifford, B. (2005) 'Development and the Environment: Perception and Opinion in St Andrews, Scotland', *Scottish Geographical Journal*, 121(4), pp. 355–384.

Webb, S. (2020) 'Engagements, Conservatives', https://www.suzannewebb.org.uk/news/prime-ministers-questions-protecting-green-belt (Accessed: 4/1/2024).

Welsh, I. and Wynne, B. (2013) 'Science, Scientism and Imaginaries of Publics in the UK: Passive Objects, Incipient Threats', *Science as Culture*, 22(4), pp. 540–566.

West Midlands Group (1948) *Conurbation: A Survey of Birmingham and the Black Country* (London: The Architectural Press).

Wetzstein, S. (2017) 'The Global Urban Housing Affordability Crisis', *Urban Studies*, 54(14), pp. 3159–3177.

Whall, H. (2015) *Oxfordshire Green Belt Survey* (Oxford: CPRE).

Whitehand, J. (2001) 'British Urban Morphology: The Conzenian Tradition', *Urban Morphology*, 5(2), pp. 103–109.

Whitehead, C., Sagor, E., Edge, A. and Walker, B. (2015) *Understanding the Local Impact of New Residential Development: A Pilot Study* (London: LSE).

Wilding, M. (2018) 'How We Did It: Planning 7,000 Homes in the Green Belt', *Planning Resource*, https://www.planningresource.co.uk/article/1495032/it-planning-7000-homes-green-belt (Accessed: 7/8/2024).

Wills, J. (2016) 'Emerging Geographies of English Localism: The Case of Neighbourhood Planning', *Political Geography*, 53, pp. 43–53.

Wilson, A., Tewdwr-Jones, M. and Comber, R. (2019) 'Urban Planning, Public Participation and Digital Technology: App Development as a Method of Generating Citizen Involvement in Local Planning Processes', *Environment and Planning B: Urban Analytics and City Science*, 46(2), pp. 286–302.

Winchester, H. (1999) 'Interviews and Questionnaires as Mixed Methods in Population Geography: The Case of Lone Fathers in Newcastle, Australia', *The Professional Geographer*, 51(1), pp. 60–67.

Wolf, M. (2015B) The Solution to England's Housing Crisis Lies in the Green Belt, *Financial Times*, http://www.ft.com/cms/s/0/f5b26d8a-ac59-11e4-9d32-00144feab7de.html#axzz3ua73wCNr (Accessed: 6/2/2024).

Wolf, M. (2015A) A Radical Solution for England's Housing Crisis, *Financial Times*, http://www.ft.com/cms/s/0/f8cdae58-9f2e-11e5-8613-08e211ea5317.html#axzz3ua73wCNr (Accessed: 6/2/2023).

Woods, M. and Gardner, G. (2011) 'Applied Policy Research and Critical Human Geography: Some Reflections on Swimming in Murky Waters', *Dialogues in Human Geography*, 1(2), pp. 198–214.

Yarwood, R. (2002) 'Parish Councils, Partnership and Governance: The Development of "exceptions" Housing in the Malvern Hills District, England', *Journal of Rural Studies*, 18(3), pp. 275–291.

Yeung, H. (1997) 'Critical Realism and Realist Research in Human Geography: A Method or a Philosophy in Search of a Method?', *Progress in Human Geography*, 21(1), pp. 51–74.

Yin, R. (2017) *Case Study Research: Design and Methods* (London: Sage).

Yokohari, M., Takeuchi, K., Watanabe, T. and Yokota, S. (2000) 'Beyond Greenbelts and Zoning: A New Planning Concept for the Environment of Asian Mega-Cities', *Urban Ecology: An International Perspective on the Interaction Between Humans and Nature*, 47, pp. 783–796.

Yorke, H. and Neilan, C. (2021) 'Liberal Democrats Take Tory Stronghold With Historic Win in Chesham and Amersham By-election', *The Telegraph*, https://www.telegraph.co.uk/politics/2021/06/18/liberal-democrats-take-tory-stronghold-historic-win-chesham/ (Accessed 03/04/2024).

YouGov (2015) 'Broadway Malyan/YouGov Survey Results', https://d25d2506sfb94s.cloudfront.net/cumulus_uploads/document/p5w7yb920c/BroadwayMalyanResults_150407_housing_Website.pdf (Accessed: 10/7/2024).

YouGov (2018) 'Survey Results', https://d25d2506sfb94s.cloudfront.net/cumulus_uploads/document/rljpph9ohs/HousingQs_FullTables_W.pdf (Accessed: 7/5/2025).

Young, C. (2019) 'Green Belt: Time For A Rethink', *LinkedIn*, https://www.linkedin.com/pulse/green-belt-time-rethink-christopher-young-qc/ (Accessed 05/04/2024).

Young, C. (2020) 'Explained: Planning White Paper and 'Current Changes'', *LinkedIn*, https://www.linkedin.com/feed/update/urn:li:activity:6709337597438672897/(Accessed: 9/10/2023).

Young, C. (2021) Planning White Paper Explained: Paper 2 (No. 5 Chambers: Birmingham).

Young, C. (2023) NPPF Consultation 2023 (No. 5 Chambers: Birmingham).

Zonneveld, W. (2007) 'A Sea of Houses: Preserving Open Space in an Urbanised Country', *Journal of Environmental Planning and Management*, 50(5), pp. 657–675.

Index

For Product Safety Concerns and Information please contact our EU
representative GPSR@taylorandfrancis.com
Taylor & Francis Verlag GmbH, Kaufingerstraße 24, 80331 München, Germany

www.ingramcontent.com/pod-product-compliance
Ingram Content Group UK Ltd.
Pitfield, Milton Keynes, MK11 3LW, UK
UKHW021519300425
457996UK00007B/30